CONFRONTING GLOBAL WARMING

The Role
of Industry

CONFRONTING GLOBAL WARMING

The Role of Industry

Tom Streissguth

Michael E. Mann
Consulting Editor

GREENHAVEN PRESS
A part of Gale, Cengage Learning

Detroit • New York • San Francisco • New Haven, Conn • Waterville, Maine • London

GALE
CENGAGE Learning™

Christine Nasso, *Publisher*
Elizabeth Des Chenes, *Managing Editor*

© 2011 Greenhaven Press, a part of Gale, Cengage Learning

For more information, contact:

Greenhaven Press
27500 Drake Rd.
Farmington Hills, MI 48331-3535
Or you can visit our Internet site at
gale.cengage.com.

For product information and technology assistance, contact us at
Gale Customer Support, 1-800-877-4253.
For permission to use material from this text or product, submit all requests online at
www.cengage.com/permissions.
Further permissions questions can be e-mailed to
permissionrequest@cengage.com

Every effort is made to ensure that Greenhaven Press accurately reflects the original intent of the authors. Every effort has been made to trace the owners of copyrighted material.

Cover image © Charles O'Rear/Corbis. Leaf image copyright © JinYoung Lee, 2010, used under license from Shutterstock.com. Leaf icon © iStockPhoto.com/domin_domin.

**LIBRARY OF CONGRESS
CATALOGING-IN-PUBLICATION DATA**
Streissguth, Thomas, 1958-
 The role of industry / by Tom Streissguth.
  p. cm. -- (Confronting global warming)
 Includes bibliographical references and index.
 ISBN 978-0-7377-5176-5 (hardcover)
 1. Industrial revolution. 2. Global warming. 3. Clean energy industries. I. Title.
 HD2328.S77 2011
 363.738'741--dc22
                              2011005924

Printed in the United States of America
1 2 3 4 5 6 7 15 14 13 12 11

Contents

Preface

> *"The warnings about global warming have been extremely clear for a long time. We are facing a global climate crisis. It is deepening. We are entering a period of consequences."*
> *Al Gore*

Still hotly debated by some, human-induced global warming is now accepted in the scientific community. Earth's average yearly temperature is getting steadily warmer; sea levels are rising due to melting ice caps; and the resulting impact on ocean life, wildlife, and human life is already evident. The human-induced buildup of greenhouse gases in the atmosphere poses serious and diverse threats to life on earth. As scientists work to develop accurate models to predict the future impact of global warming, researchers, policy makers, and industry leaders are coming to terms with what can be done today to halt and reverse the human contributions to global climate change.

Each volume in the Confronting Global Warming series examines the current and impending challenges the planet faces because of global warming. Several titles focus on a particular aspect of life—such as weather, farming, health, or nature and wildlife—that has been altered by climate change. Consulting the works of leading experts in the field, Confronting Global Warming authors present the current status of those aspects as they have been affected by global warming, highlight key future challenges, examine potential solutions for dealing with the results of climate change, and address the pros and cons of imminent changes and challenges. Other volumes in the series—such as those dedicated to the role of government, the role of industry, and the role of the individual—address the impact various facets of society can have on climate change. The result is a series that provides students and general-interest readers with a solid understanding of the worldwide ramifications of climate change and what can be done to help humanity adapt to changing conditions and mitigate damage.

Each volume includes:

- A descriptive **table of contents** listing subtopics, charts, graphs, maps, and sidebars included in each chapter
- Full-color **charts, graphs, and maps** to illustrate key points, concepts, and theories
- Full-color **photos** that enhance textual material
- **Sidebars** that provide explanations of technical concepts or statistical information, present case studies to illustrate the international impact of global warming, or offer excerpts from primary and secondary documents
- **Pulled quotes** containing key points and statistical figures
- A **glossary** providing users with definitions of important terms
- An annotated **bibliography** of additional books, periodicals, and websites for further research
- A detailed **subject index** to allow users to quickly find the information they need

The Confronting Global Warming series provides students and general-interest readers with the information they need to understand the complex issue of climate change. Titles in the series offer users a well-rounded view of global warming, presented in an engaging format. Confronting Global Warming not only provides context for how society has dealt with climate change thus far but also encapsulates debates about how it will confront issues related to climate in the future.

Foreword

Earth's climate is a complex system of interacting natural components. These components include the atmosphere, the ocean, and the continental ice sheets. Living things on earth—or, the biosphere—also constitute an important component of the climate system.

Natural Factors Cause Some of Earth's Warming and Cooling

Numerous factors influence Earth's climate system, some of them natural. For example, the slow drift of continents that takes place over millions of years, a process known as plate tectonics, influences the composition of the atmosphere through its impact on volcanic activity and surface erosion. Another significant factor involves naturally occurring gases in the atmosphere, known as greenhouse gases, which have a warming influence on Earth's surface. Scientists have known about this warming effect for nearly two centuries: These gases absorb outgoing heat energy and direct it back toward the surface. In the absence of this natural greenhouse effect, Earth would be a frozen, and most likely lifeless, planet.

Another natural factor affecting Earth's climate—this one measured on timescales of several millennia—involves cyclical variations in the geometry of Earth's orbit around the sun. These variations alter the distribution of solar radiation over the surface of Earth and are responsible for the coming and going of the ice ages every one hundred thousand years or so. In addition, small variations in the brightness of the sun drive minor changes in Earth's surface temperature over decades and centuries. Explosive volcanic activity, such as the Mount Pinatubo eruption in the Philippines in 1991, also affects Earth's climate. These eruptions inject highly reflective particles called aerosol into the upper part of the atmosphere, known as the stratosphere, where they can reside for a year or longer. These particles reflect some of the incoming sunlight back into space and cool Earth's surface for years at a time.

Human Progress Puts Pressure on Natural Climate Patterns

Since the dawn of the industrial revolution some two centuries ago, however, humans have become the principal drivers of climate change. The burning of fossil fuels—such as oil, coal, and natural gas—has led to an increase in atmospheric levels of carbon dioxide, a powerful greenhouse gas. And farming practices have led to increased atmospheric levels of methane, another potent greenhouse gas. If humanity continues such activities at the current rate through the end of this century, the concentrations of greenhouse gases in the atmosphere will be higher than they have been for tens of millions of years. It is the unprecedented rate at which we are amplifying the greenhouse effect, warming Earth's surface, and modifying our climate that causes scientists so much concern.

The Role of Scientists in Climate Observation and Projection

Scientists study Earth's climate not just from observation but also from a theoretical perspective. Modern-day climate models successfully reproduce the key features of Earth's climate, including the variations in wind patterns around the globe, the major ocean current systems such as the Gulf Stream, and the seasonal changes in temperature and rainfall associated with Earth's annual revolution around the sun. The models also reproduce some of the more complex natural oscillations of the climate system. Just as the atmosphere displays random day-to-day variability that we term "weather," the climate system produces its own random variations, on timescales of years. One important example is the phenomenon called El Niño, a periodic warming of the eastern tropical Pacific Ocean surface that influences seasonal patterns of temperature and rainfall around the globe. The ability to use models to reproduce the climate's complicated natural oscillatory behavior gives scientists increased confidence that these models are up to the task of mimicking the climate system's response to human impacts.

To that end, scientists have subjected climate models to a number of rigorous tests of their reliability. James Hansen of the NASA Goddard Institute for Space Studies performed a famous experiment back in 1988, when he subjected a climate model (one relatively primitive by modern standards) to possible future fossil fuel emissions scenarios. For the scenario that most closely matches actual emissions since then, the model's predicted course of global temperature increase shows an uncanny correspondence to the actual increase in temperature over the intervening two decades. When Mount Pinatubo erupted in the Philippines in 1991, Hansen performed another famous experiment. Before the volcanic aerosol had an opportunity to influence the climate (it takes several months to spread globally throughout the atmosphere), he took the same climate model and subjected it to the estimated atmospheric aerosol distribution. Over the next two years, actual global average surface temperatures proceeded to cool a little less than 1°C (1.8°F), just as Hansen's model predicted they would.

Given that there is good reason to trust the models, scientists can use them to answer important questions about climate change. One such question weighs the human factors against the natural factors to determine responsibility for the dramatic changes currently taking place in our climate. When driven by natural factors alone, climate models do not reproduce the observed warming of the past century. Only when these models are also driven by human factors—primarily, the increase in greenhouse gas concentrations—do they reproduce the observed warming. Of course, the models are not used just to look at the past. To make projections of future climate change, climate scientists consider various possible scenarios or pathways of future human activity.

The earth has warmed roughly 1°C since preindustrial times. In the "business as usual" scenario, where we continue the current course of burning fossil fuel through the twenty-first century, models predict an additional warming anywhere from roughly 2°C to 5°C (3.6°F to 9°F). The models also show that even if we were to stop fossil fuel burning today, we are probably committed to as much as 0.6°C additional warming because of the inertia of the climate system. This inertia ensures warming for a century to come, simply due to our greenhouse gas emissions thus far. This committed warming introduces a profound procrastination penalty for not taking immediate action. If we are to avert an additional warming of 1°C, which would bring the net warming to 2°C—often considered an appropriate threshold for defining dangerous human impact on our climate—we have to act almost immediately.

Long-Term Warming May Bring About Extreme Changes Worldwide

In the "business as usual" emissions scenario, climate change will have an array of substantial impacts on our society and the environment by the end of this century. Patterns of rainfall and drought are projected to shift in such a way that some regions currently stressed for water resources, such as the desert southwest of the United States and the Middle East, are likely to become drier. More intense rainfall events in other regions, such as Europe and the midwestern United States, could lead to increased flooding. Heat waves like the one in Europe in summer 2003, which killed more than thirty thousand people, are projected to become far more common. Atlantic hurricanes are likely to reach greater intensities, potentially doing far more damage to coastal infrastructure.

Furthermore, regions such as the Arctic are expected to warm faster than the rest of the globe. Disappearing Arctic sea ice already threatens wildlife, including polar bears and walruses. Given another 2°C warming (3.6°F), a substantial portion of the

Greenland ice sheet is likely to melt. This event, combined with other factors, could lead to more than 1 meter (about 3 feet) of sea-level rise by the end of the century. Such a rise in sea level would threaten many American East Coast and Gulf Coast cities, as well as low-lying coastal regions and islands around the world. Food production in tropical regions, already insufficient to meet the needs of some populations, will probably decrease with future warming. The incidence of infectious disease is expected to increase in higher elevations and in latitudes with warming temperatures. In short, the impacts of future climate change are likely to have a devastating impact on society and our environment in the absence of intervention.

Strategies for Confronting Climate Change

Options for dealing with the threats of climate change include both adaptation to inevitable changes and mitigation, or lessening, of those changes that we can still affect. One possible adaptation would be to adjust our agricultural practices to the changing regional patterns of temperature and rainfall. Another would be to build coastal defenses against the inundation from sea-level rise. Only mitigation, however, can prevent the most threatening changes. One means of mitigation that has been given much recent attention is geoengineering. This method involves perturbing the climate system in such a way as to partly or fully offset the warming impact of rising greenhouse gas concentrations. One geoengineering approach involves periodically shooting aerosol particles, similar to ones produced by volcanic eruptions, into the stratosphere—essentially emulating the cooling impact of a major volcanic eruption on an ongoing basis. As with nearly all geoengineering proposals, there are potential perils with this scheme, including an increased tendency for continental drought and the acceleration of stratospheric ozone depletion.

The only foolproof strategy for climate change mitigation is the decrease of greenhouse gas emissions. If we are to avert a dangerous 2°C increase relative to preindustrial times, we will probably need to bring greenhouse gas emissions to a peak within the coming years and reduce them well below current levels within the coming decades. Any strategy for such a reduction of emissions must be international and multipronged, involving greater conservation of energy resources; a shift toward alternative, carbon-free sources of energy; and a coordinated set of governmental policies that encourage responsible corporate and individual practices. Some contrarian voices argue that we cannot afford to take such steps. Actually, given the procrastination penalty of not acting on the climate change problem, what we truly cannot afford is to delay action.

Evidently, the problem of climate change crosses multiple disciplinary boundaries and involves the physical, biological, and social sciences. As an issue facing all of civilization, climate change demands political, economic, and ethical considerations. With the Confronting Global Warming series, Greenhaven Press addresses all of these considerations in an accessible format. In ten thorough volumes, the series covers the full range of climate change impacts (water and ice; extreme weather; population, resources, and conflict; nature and wildlife; farming and food supply; health and disease) and the various essential components of any solution to the climate change problem (energy production and alternative energy; the role of government; the role of industry; and the role of the individual). It is my hope and expectation that this series will become a useful resource for anyone who is curious about not only the nature of the problem but also about what we can do to solve it.

Michael E. Mann

Michael E. Mann is a professor in the Department of Meteorology at Penn State University and director of the Penn State Earth System

Science Center. In 2002 he was selected as one of the fifty lead-ing visionaries in science and technology by Scientific American. *He was a lead author for the "Observed Climate Variability and Change" chapter of the Intergovernmental Panel on Climate Change (IPCC) Third Scientific Assessment Report, and in 2007 he shared the Nobel Peace Prize with other IPCC authors. He is the author of more than 120 peer-reviewed publications, and he recently coauthored the book* Dire Predictions: Understanding Global Warming *with colleague Lee Kump. Mann is also a co-founder and avid contributor to the award-winning science Web site RealClimate.org.*

Chapter 1

The Industrial Revolution and Global Warming

Europe's industrial revolution began in eighteenth-century Britain. The English textile industry, once reliant on hand-operated looms, adopted new machinery that allowed workers to mass-produce clothing. Factories employed thousands of workers to operate looms and spinning machines. The textile industry thrived as wool and cotton cloth produced by these machines was exported throughout Europe and the British colonies.

The industrial revolution changed the nature of work. Instead of laboriously making cloth at home, textile workers were gathered in large factories designed for mass production. Promising steady employment and wages, these factories drew rural farmers and laborers into the cities. By the end of the eighteenth century, when industrialization was in full progress, London held half the people in England and was the most populous city in Europe.

Textile companies in England competed with similar industries operating in the colonies. Firms in the parent country held enough clout in the British parliament to have their competitors closed down. Britain also protected its domestic industries by levying duties and taxes on goods imported from abroad. But there was little government regulation of working conditions or wages, and no limits were placed on the damage done by factory emissions and waste. The natural environment was of little concern to the king or his royal officials, or to parliament, who all had a much greater interest in promoting the success of English industry.

Coal-Fired Steam Engines

After inventor James Watt improved the steam engine, machines began to take over the tedious and time-consuming work of making goods. At the same time, England's transportation system was going through a steam revolution. The use of canals and boats to transport goods throughout the country gradually gave way to railroads and steam-powered trains. Small tramways had been used by coal miners to move their hauling wagons since the 1700s. During the 1820s inventor George Stephenson built the first freight-and-passenger railroad, the Stockton and Darlington Railway.

The most efficient fuel for driving steam locomotives was coal. This fuel burned hotter and with less smoke than wood, and it was much more plentiful. England had sources of coal in the northern counties where textiles had been manually produced for centuries. In North America, coal mining began in the Appalachian Mountains of Virginia during the 1740s. Beginning in the nineteenth century, steam engines using coal-fired boilers powered ships and railway locomotives. Coal-fired machines could also generate the high temperatures needed to produce iron and steel.

Coal mines remained dangerous, unhealthy places to work. Miners labored in narrow tunnels subject to collapse. They risked death by suffocation, by explosion, and by fire. Favored for their small size, children were employed to haul coal to the surface and otherwise assist the miners. Coal mines destroyed hillsides and vegetation, and the use of coal-fired steam engines further polluted England's gray skies.

Following pages: Coal-fired boilers began to power train locomotives in the nineteenth century. This Thatcher Perkins steam engine (#147) was built in 1863. The coal was carried in a tender car behind the engine. Robert Nickelsberg/Getty Images.

Steel and Iron Production

Since the seventeenth century, the iron industry in England had been using coal to fuel forges and furnaces to transform iron ore into useful metal. New blast furnaces of the 1700s allowed producers to make tools, swords, and kitchen utensils that were cheaper, lighter, and easier to use. By the 1800s, the industry was using charcoal made from abundant coal (a by-product called coke) rather than scarcer wood charcoal as its primary fuel. This shift helped, somewhat, to slow the deforestation of the English countryside.

> *The first air-quality regulation in England dates to 1306, when King Edward I attempted to ban coal burning. Despite the threatened punishment of death, the ban was widely ignored, and air quality in many English towns remained poor.*

England exported its iron products throughout Europe and to its vast global colonial empire. Iron went into modern bridges, which engineers could design to cross much greater spans than before, as well as into the rails necessary for the expanding railroad network. The development of new methods of steelmaking also brought down the cost of important commodities. Cheap iron and steel allowed factories to mass-produce the boilers and steam engines needed for manufacturing.

But steam-fired boilers using coal as an energy source contributed to the smoggy skies over London and other English cities. Smoke, dust, and debris clogged the air, worsening the unhealthy conditions in which urban factory laborers lived.

Foggy Skies: Pollution of Early Industrial Cities

The first air-quality regulation in England dates to 1306, when King Edward I attempted to ban coal burning. Despite the threat-

Eyewitness to Smog

Although the people of nineteenth-century England grew accustomed to the poor air created by coal-fired steam engines and home fireplaces, the "Fog" often came as an unpleasant shock to visitors. In her London Journal *(originally written in 1840), traveler Flora Tristan commented unfavorably on the state of the air in London and other English cities.*

Over every English town there hangs a pall compounded of the Ocean vapours that perpetually shroud the British Isles, and the heavy noxious fumes of the Cyclops' cave. No longer does timber from the forests provide fuel for the family hearth; the fuel of Hell, snatched from the very bowels of the earth, has usurped its place. It burns everywhere, feeding countless furnaces, replacing horsepower on the roads and wind-power on the rivers and the seas which surround the empire.

Above the monster city a dense fog combines with the volume of smoke and soot issuing from thousands of chimneys to wrap London in a black cloud which allows only the dimmest light to penetrate and shrouds everything in a funeral veil.

In London melancholy is in the very air you breathe and enters in at every pore. There is nothing more gloomy or disquieting than the aspect of the city on a day of fog or rain or black frost. Only succumb to its influence and your head becomes painfully heavy, your digestion sluggish, your respiration laboured for lack of fresh air, and your whole body is overcome by lassitude. Then you are in the grip of what the English call "spleen": a profound despair, unaccountable anguish, cantankerous hatred for those one loves the best, disgust with everything, and an irresistible desire to end one's life by suicide.

SOURCE: *The London Journal of Flora Tristan*, London: Virago, 1982, p. 22.

ened punishment of death, the ban was widely ignored, and air quality in many English towns remained poor. Centuries later, in the United States, polluted air and water were commonplace in the industrial East and Midwest. In Chicago, dark smoke poured from factory chimneys, railway locomotives, and mills. The vast slaughterhouses on the city's south side spread an infamous stench, especially on warm and humid days.

To address the problem of air pollution, in 1892 a group of Chicago businessmen formed the Society for Prevention of Smoke. The society persuaded the city of Chicago to pass a nuisance ordinance against boats that regularly emitted thick clouds of black smoke while traveling the Chicago River. The law was feebly enforced and also controversial, as it seemed to favor the anthracite coal that had to be imported from the east (coal mined in Illinois was of the smoky, bituminous variety). During the nineteenth century, coal smoke also prompted the passage of an antipollution ordinance in St. Louis.

A Revolution in Daily Life: Food, Vehicles, Goods

In 1879 Thomas Edison began improvements upon the incandescent bulb, which English and German inventors had been developing since the early nineteenth century. The light bulb would ultimately enable people to light their homes with direct-current electricity rather than with oil or gas lamps. By 1888 Nikola Tesla had invented a system of power distribution using alternating electrical current. This advance allowed electricity to be sent over a network of transmission lines and to be used as a source of power by factories, homes, and streetcars. Over the next few decades, hundreds of new electrical devices were invented for the convenience of households and offices.

The first electricity companies relied on hydroelectric plants to generate current, but over the years coal-fired steam turbines, which don't depend on a controlled flow of water, became more widespread. According to the Union of Concerned Scientists, 54 percent of all electricity produced in the United States is still generated by coal-fired plants.[1]

The Internal Combustion Engine

The use of fossil fuels grew rapidly with the invention of the internal combustion engine and the mass-marketing of automobiles in the early twentieth century. Pound for pound, oil-based petroleum had a higher energy content than coal. Liquid fuels were easier to store, and extracting them from the earth did less harm to the environment. To this day, most vehicles still depend on gasoline and other forms of refined crude oil. Liquid fuel is

The Donora Smog

A poisonous yellow cloud of smoke and grit rolled into the streets of Donora, Pennsylvania, on October 27, 1948. In the homes of this small town in the Monongahela River valley, thousands of people began to suffer coughing fits. The thick cloud made it difficult to drive, and many people going outdoors, even for short periods of time, suffered dizziness, headaches, and chest pains.

In Donora, the US Steel Corporation operated the American Steel and Wire plant and the Donora Zinc Works, the largest zinc mill in the world. These two factories, which worked seven days a week and around the clock, regularly emitted a toxic mix of pollutants, including fatal levels of fluorine.

Over the next four days, the smog lay heavy over Donora, trapped by a temperature inversion that prevented circulation of the air out of the valley. Twenty people and several hundred house pets died of respiratory distress, and as many as seven thousand people—half the population—had taken sick. On October 31, a Sunday, US Steel began shutting down the plants, which remained closed throughout the day. On Monday morning, after rain dispersed the toxic cloud, the plants resumed operation.

The Donora Smog, as it became known, was an important event in the modern environmental movement. To people in Donora, the Monongahela valley, and the rest of the country, the dangers of industrial pollution became apparent and the event inspired a drive for federal regulation of industrial emissions.

also used to power jet aircraft, for diesel engines used in factories, to heat homes, and to generate electricity.

After the end of World War I, the United States went through some hard economic times. Many businesses failed as the work generated by war production disappeared. Jobs were hard to find, and race riots and other civil trouble shook the nation. But boom times returned in the 1920s. Factories produced automobiles and household goods. Banks and stores freely made loans for the purchase of homes, cars, furniture, and appliances. Workers became consumers, striving for a more comfortable and convenient life through the use of gadgets powered by electricity. The demand for energy rose with the new consumer culture.

In the industrial East, coal-fired power plants allowed factories to work around the clock. The bituminous coal dug from the Appalachian Mountains was high in energy content, measured in BTUs (British thermal units), but was also heavily polluting. Burning this coal emitted into the atmosphere a heavy stream of smoke and soot that contained carbon dioxide (CO_2), mercury, and other toxic pollutants.

For the people of Pittsburgh, the smoke and dust from coal burning was just an everyday inconvenience. Air pollution symbolized the city's prosperity, which relied on heavy industries such as steelmaking. The city placed few limits on factory emissions. The Bureau of Smoke Control had only weak enforcement authority. Factories did not have the technology to treat their emissions before pollutants escaped their smokestacks. Nobody gave a thought to the greenhouse effect, the contribution of carbon dioxide and other gases to the warming of the atmosphere.

Yet other cities, including St. Louis and Chicago, had managed to improve their air quality by passing ordinances limiting factory emissions. In 1946, Pittsburgh followed suit with the enactment of its first smoke-control law, aimed at coal burning by steel plants and locomotives, as well as by individual homes. Many Pittsburgh factories complied by converting their machinery from coal to natural gas, a fuel brought north via pipeline from gas fields in Texas and Oklahoma. In the meantime, the Pennsylvania Railroad and other rail companies built diesel engines to replace coal-burning locomotives.

The Great London Pea Souper Leads to Efforts Toward Clean Air

The city of London was famous for its fogs and its "pea soupers," fog events that lasted several days and covered the city in a thick shroud of grayish, smokey air. London's fogs were caused by temperature inversions, in which heavy, colder air aloft prevented the escape of warmer, polluted air near the surface.

One especially thick pea souper attained legendary status among Londoners. The "Big Smoke" took place during several windless, cold days in early December 1952. As millions of households used coal fires to stay warm, smoke and soot covered the city, trapped by a dense layer of cold air that would not rise or disperse. For those citizens bold enough to venture outside, poisonous concentrations of sulfur dioxide and particulate matter made it painful to breathe. Visibility was so bad that train journeys and aircraft flights were cancelled. Several thousand people died from bronchitis, heart attacks, or pneumonia.

The Big Smoke resulted in a drive among English lawmakers to clean up London's air and in the passage of the 1956 Clean Air Act. The government provided money to help residents throughout England convert their coal-burning heaters to gas, oil, or electricity. This law was followed in 1968 by another Clean Air Act, which required factories and mills to use taller chimneys to vent fumes and smoke. Although this measure did release pollutants at a higher elevation, it also resulted in pollution clouds that reached, in some cases, thousands of miles from their point of origin.

While the English government began regulating pollution, the United States still had no federal (national) laws limiting emissions or setting air quality standards. Toxic air was considered a local matter, and before 1970 state, city, and county governments were deemed responsible for passing pollution ordinances as needed. In some places, environmental regulation had been a success. St. Louis, Chicago, and Pittsburgh had all improved their air with restrictions on emissions. In these cities, soot-covered buildings were cleaned, the rate of respiratory diseases caused by foul air decreased, and the overall quality of life improved.

But as factory smoke diminished, automobile exhaust intensified. Los Angeles had few large factories, but the city still suffered a brown haze that grew thicker and darker each year. The new and growing car culture of the United States—and other nations—was having a visible effect on the air in Los Angeles and other large cities around the world.

Industrialization Is Correlated to a Rise in Carbon Dioxide Emissions

In 1952 Arie Haagen-Smit made an important discovery about automobile emissions and the atmosphere: The brown haze that was rising above Los Angeles and other major cities was a form of photochemical smog. Haagen-Smit found that when ultraviolet radiation from the sun strikes automobile exhaust, a chemical reaction occurs. Smog and ozone result, turning the air brown and unhealthful.

In 1959, the state of California passed a new law that established air-quality standards. The law also required emission control devices on automobiles. The Department of Public Health set standards for the concentrations of sulfur dioxide, nitrogen dioxide, carbon monoxide, suspended particulates, and photochemical oxidants. These environmental regulations were the first of many passed by California that later served as models for national standards.

In the meantime, scientists were studying the effect of carbon dioxide emissions, which are created whenever fossil fuels such as wood, oil, or coal are burned. Svante Arrhenius, a Swedish scientist, had been the first researcher to correlate rising atmospheric concentrations of carbon dioxide with rising temperatures. Arrhenius believed that adding more carbon dioxide through fossil-fuel burning would contribute to a greenhouse effect, in which this gas traps some solar radiation (heat) in the atmosphere before the radiation escapes into space.

For several decades, scientists dismissed Arrhenius's theory and the greenhouse effect. The earth and its atmosphere were so vast, it was believed, that human activity could not possibly have any significant effect on climate. Although coal burning did contribute to pollution in some places, an anthropogenic (human-caused) global warming effect seemed a far-fetched concept.

More research in the 1950s revealed some important and relevant facts, however. Improved measuring instruments showed that carbon dioxide absorbs infrared radiation in the atmosphere. It was also found that carbon dioxide remains in the atmosphere about ten years after being released.[2] Scientists also debated a hypothesis that the ocean serves as a carbon sink (absorber). Although it was known that the ocean did absorb carbon dioxide, scientists were unsure of the rate of that absorption or of the effect on marine life of gradually rising CO_2 levels.

The Keeling Curve

In 1955 the United States government passed the Air Pollution Control Act, then in 1963 the Clean Air Act. These federal laws did not regulate pollution but did authorize research into the sources and extent of air pollution. Car emissions were first controlled by the Clean Air Act of 1970. This law defined hazardous air pollutants and established the authority of the federal government to control the amount that could be put into the air.

Atmospheric researchers were investigating Svante Arrhenius's hypothesis that rising levels of carbon dioxide would cause a rise in global temperatures. To determine the accuracy of this hypothesis, the American researcher Charles Keeling ventured up the slopes of Mauna Loa, a volcanic summit on the island of Hawaii, in 1958. On Mauna Loa, Keeling installed equipment

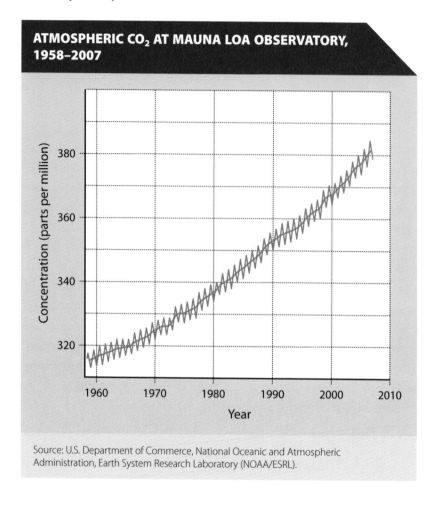

ATMOSPHERIC CO₂ AT MAUNA LOA OBSERVATORY, 1958–2007

Source: U.S. Department of Commerce, National Oceanic and Atmospheric Administration, Earth System Research Laboratory (NOAA/ESRL).

that measures carbon dioxide concentration in the atmosphere. The chart that resulted from these measurements showed clearly that CO_2 emissions were on the rise each year. They had reached levels well above the historical norm of 280 parts per million, a measurement that scientists had obtained through the analysis of ancient ice cores.

By 2010, the Keeling measurement had reached 380 parts per million, more than 30 percent higher than the historic average, and was rising at an accelerating rate.[3] As the concentration of carbon dioxide continues to rise, and climatologists continue to record the slow but steady rise in global average temperatures, nations around the world are debating the best course of action to slow the warming trend.

The basic controversy revolves around means, not ends. Either governments can set carbon emission limits, and attempt to force industries to comply through taxes and fines, or they can allow business markets to work through a system of trading emission allowances, which encourage conservation of energy and the use of energy sources that do not contribute to greenhouse gas levels.

The debate has caused friction between developing countries (that are seeking to grow their economies as quickly as possible) and developed nations (that want to maintain their higher living standards). The role of industry in this controversy is crucial to its eventual resolution; energy generation, transportation industries, and manufacturing are the world's most significant sources of greenhouse gases. Without the cooperation, either forced or voluntary, of large industries, the issue of how best to address global warming will never be resolved.

Notes

1. Union of Concerned Scientists, "Coal vs. Wind." www.ucsusa.org.
2. US Department of Energy, "A Brief History of Coal Use." http://fossil.energy.gov.
3. Earth System Research Laboratory, "CCGG Observatory Measurements." www.esrl.noaa.gov.

CHAPTER 2

Industrial Emissions

Together, the world's manufacturing industries emit about one third of all carbon dioxide (CO_2) released into the earth's atmosphere. Among them are primary industries such as chemical, cement, steel, and paper factories that need large amounts of energy for powering assembly lines, generators, and other machinery. In fast-developing countries such as India and China, the growth of primary industries has brought a sharply rising level of greenhouse gas emissions. Older factories throughout the world rely on obsolete machinery and use inefficient energy sources such as high-sulfur coal for power production.

Businesses strive for efficiency and lowered costs. As they compete with other companies, they seek out less expensive raw materials as well as cheaper energy. The technology for more efficient use of energy is widely available; the problem, for most companies, is the high cost of investment in new plants and equipment. In developing countries, the capital needed for such investment may be difficult to come by, as many of these nations depend heavily on foreign investment for business development.

The story is different in developed nations in North America, Europe, and Asia. According to a report titled "Japan: A Model of Energy Efficiency?" the most energy-efficient factories in the world operate in East Asia, particularly Japan. The Japanese began taking energy-saving measures during the 1970s as a result of oil-price hikes. In the report, economist Yukari Yamashita comments that "The prices of everything went up because of the oil crisis, so everybody was aware that we have to do something, otherwise our life won't be sustainable." The report continues: "The [Japanese] government held emergency meetings, quickly passing a series of conservation laws, forcing factories to replace old, inefficient boilers and assembly-line machinery with new energy-saving equipment. Rising energy taxes funded programs like low-interest loans for companies developing more energy-efficient solutions for industry."[1]

In places where industry has received outside investment, energy efficiency also improves. New aluminum smelters built by Chinese firms in Africa, for example, are among the most efficient and greenest in the world. Aluminum production accounts for nearly half the demand for electricity among all nonferrous metal industries, and the cost of electricity is an important consideration for new factories in this business. A 2008 report published by the International Energy Agency reveals that "New smelters tend to be based on the latest technology and energy efficiency is a key consideration in smelter development."[2]

While energy efficiency helps factories produce goods at lower cost, it also helps reduce greenhouse gases. One study by the International Energy Agency found that improved energy efficiency offers a potential reduction of 7 to 12 percent of carbon dioxide emissions worldwide.[3]

There are various options open to factories seeking to achieve greater efficiency and lower costs, and to reduce the emission of greenhouse gases. The technologies available include adjustable speed drives, which can power assembly lines at different rates of speed. An adjustable speed drive saves money and energy when it replaces a set of fixed-speed drives that are constantly cycled on and off during the manufacturing process. Combined heat and power allows factories to recycle their emissions to generate heat and electricity. New recycling technology also helps factories use raw materials made from waste products, such as plastics and rubber, either as feedstock for their finished goods or as fuel for their generators.

In recent decades, most manufacturing industries have become more energy efficient as new plants with new technology are built to replace older ones. A trend toward construction of larger plants, which are more efficient than smaller ones, also helps in the effort. Carbon-trading mechanisms allow companies to buy and trade permits for the emission of a certain amount of carbon and give companies an incentive to invest in energy-efficient technologies.

Although occasional government regulation of emissions dates back to the fourteenth century and the decrees of King Edward I, England and other countries paid little heed to environmental issues until the industrial revolution. As manufacturing industries expanded, their operations began to make a noticeable impact on the air and water. The first uniform government regulations of industry came about through a problem first researched in the nineteenth century: acid rain.

Coal Burning and Acid Rain

With the industrial revolution came acid rain and the damage it caused to buildings and monuments in cities throughout Great Britain. An English researcher, Robert Angus Smith, discovered the source of acid rain in 1852. The release of sulfur dioxide (SO_2) into the atmosphere by coal-burning plants, Smith found, caused a chemical reaction. Water vapor mixed with the sulfur dioxide and nitrogen oxides emitted by steam-powered locomotives and coal-burning furnaces to form sulfuric acid and nitric acid. This toxic mix remained in the atmosphere, where it was driven by wind and weather systems across long distances. It returned to the surface in the form of acidic precipitation.

This polluted precipitation was a direct result of industrialization. Acid rain and acid snow falling in the northern United States and Canada damaged forests and waterways; many lakes and streams could no longer support marine life. During the

1970s, dead lakes became commonplace in the northern reaches of North America, as well as in the remote forests of Scandinavia and Russia.

Coal-burning power plants and factories are still emitting particulates and aerosols (a combination of solid material and

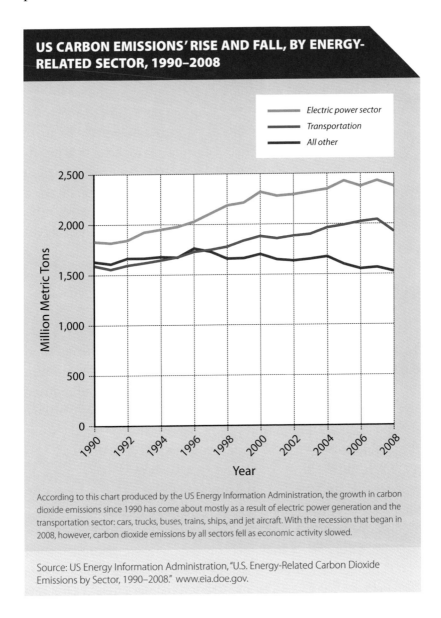

US CARBON EMISSIONS' RISE AND FALL, BY ENERGY-RELATED SECTOR, 1990–2008

Legend:
- Electric power sector
- Transportation
- All other

Y-axis: Million Metric Tons (0, 500, 1,000, 1,500, 2,000, 2,500)

X-axis: Year (1990, 1992, 1994, 1996, 1998, 2000, 2002, 2004, 2006, 2008)

According to this chart produced by the US Energy Information Administration, the growth in carbon dioxide emissions since 1990 has come about mostly as a result of electric power generation and the transportation sector: cars, trucks, buses, trains, ships, and jet aircraft. With the recession that began in 2008, however, carbon dioxide emissions by all sectors fell as economic activity slowed.

Source: US Energy Information Administration, "U.S. Energy-Related Carbon Dioxide Emissions by Sector, 1990–2008." www.eia.doe.gov.

gas). Particulate matter (dust and soot) results from artificial sources as well as natural sources such as volcanic eruptions, forest fires, and dust storms. Particulates and aerosols block sunlight and help to seed sunlight-absorbing clouds by providing tiny nuclei around which water droplets can easily form. Together, these two components have a cooling effect on the atmosphere. In high concentrations, particulates and aerosols also contribute to health problems, including respiratory diseases, heart conditions, and cancers.

Greenhouse gases and aerosols both influence the earth's climate. Whereas carbon dioxide and other gases tend to warm the surface, aerosols tend to cool it. The seesaw battle between these warming and cooling influences continues, with scientists debating whether one or the other has the more significant impact on the global climate. The key to the problem, in the opinion of many, is how long the gas and dust remain in the atmosphere. Aerosols and particulates (which are heavier than air) tend to quickly return to earth, usually in a matter of days. Carbon dioxide, on the other hand, remains suspended in the atmosphere for up to one hundred years, and thus may have a much greater influence.[4]

The Acid Rain Program

To solve the problem of sulfur dioxide emissions from coal-burning plants, the government of Canada initiated the Canadian Acid Rain Control Program in 1985. This law mandated reduced levels of sulfur dioxide emissions from Canada's factories. In the 1980s, the United States government also began to study the problem. Researchers tested lake and river water, soil, and air. They found rising concentrations of sulfur dioxide and acid compounds. In 1990, an amendment to the Clean Air Act created the Acid Rain Program.

The program capped the amount of sulfur dioxide and nitrogen oxides (NO_x) that each power plant could emit. Power plants would limit their emissions by using a new market-based system known as cap and trade. The government would set limits (or caps) on the emission of sulfur dioxide and nitrogen oxides at each plant, but the plants could also trade allowances, which were permits for the emission of the capped amount of pollutants. Plants that reduced their emissions under the limit could sell the allowances they no longer needed to other plants that were exceeding their limits. This system gave a financial incentive to companies to lower their emissions. Phase I of the program began in New England in 1995 and set a limit of 8.7 million tons of sulfur dioxide emissions for 110 of the biggest plants, to be achieved by the year 2010.

Phase II extended the program to the rest of the country. Sulfur dioxide levels gradually fell through the 1990s and the first decade of the twenty-first century. The deadline limit set in 1995 for 2010 was reached by 2007, three years ahead of schedule. In 2005, the Environmental Protection Agency established the Clean Air Interstate Rule (CAIR), which limited emissions of sulfur dioxide and nitrogen oxides throughout twenty-eight eastern states, as well as the District of Columbia. In the effort to limit damaging pollution that was traveling through the air from one state to another, CAIR applied a benchmark level reached in 2003. But acid rain that results from coal-burning plants is still a problem in northern Europe, Russia, and China.

The Growth in Emissions from Developing Nations

The fast-growing developing nations of the world, including China and India, are contributing ever-increasing amounts of industrial pollutants and carbon dioxide to the atmosphere. The issue for these nations, however, is not atmospheric pollution or the problem of global warming. Instead, the developing nations seek to expand their economies as fast as possible, provide a better standard of living for their citizens, and head off the unrest that would come with economic stagnation and joblessness.

China is growing at a fast pace, with its economy expanding by almost 10 percent each year. In 2010, Chinese production

China: Sharing the Smoke

As a nation with an economy growing at the rate of nearly 10 percent every year, China is devoting enormous energy resources to its busy manufacturing plants, which are creating mountains of consumer goods that are exported all over the world. China is giving priority to economic development over the environmental impact of burning soft brown coal, also known as lignite. This form of coal is the predominant one found in China, and to take advantage of this resource the country is building new coal-fired power plants at a rapid rate.

The pollution and carbon dioxide emissions created by use of lignite do not respect international borders, however, and China's environmental problems are now shared by distant nations. Particulate matter, an important classification of airborne pollutants, rises into the upper atmosphere and travels with the prevailing winds. Because these winds travel eastward in the northern hemisphere, China's pollution is becoming an issue for other countries, including Japan, Korea, Canada, and the United States. Hongbin Yu, a scientist researching for NASA's Goddard Space Flight Center, used satellites to track particulate matter between 2002 and 2005, reports Gretchen Cook Anderson in a 2008 article titled "NASA Satellite Measures Pollution from East Asia to North America." "Looking at four years of data from 2002 to 2005," Yu says, "we estimated the amount of pollution arriving in North America to be equivalent to about 15 percent of local emissions of the U.S. and Canada. This is a significant percentage at a time when the U.S. is trying to decrease pollution emissions to boost overall air quality. This means that any reduction in our emissions may be offset by the pollution aerosols coming from East Asia and other regions."

surpassed that of Japan, and China became the world's second-largest economy. With shifts in government policies, the private sector has rapidly developed. The economy has been boosted by investment money from Chinese exiles in Hong Kong and other places, and by foreign businesses seeking to sell their products to the largest market for goods and services in the world.

China has paid a high environmental price for rapid economic development. Acid rain affects soil and water, and as much as a third of once-productive cropland now fails to produce food. Nevertheless, China is increasing coal production and building new power plants at a rapid rate.

The vital ingredient in this economic growth is electrical energy provided by coal-fired plants, which meet about 70 percent of China's electricity demand. These plants use low-grade coal from large underground deposits that produce coal at a much lower cost than natural gas, refined crude oil, or nuclear power. The burning of coal in China, by one estimate, will emit 40 million tons of sulfur dioxide into the atmosphere by 2025. Emissions will drift over China's borders and reach the rest of eastern Asia, including Taiwan, Japan, Korea, and Thailand.[5]

China has paid a high environmental price for rapid economic development. Acid rain affects soil and water, and as much as a third of once-productive cropland now fails to produce food. Nevertheless, China is increasing coal production and building new power plants at a rapid rate. The country continues to develop its natural gas and nuclear power programs, and to adopt some environmental standards for its cities, but it gives priority to the immediate demands for light and power.

Case Study: Garbage Incineration in Mexico City

The *Distrito Federal*, or Federal District, of south-central Mexico includes Mexico City. This sprawling metropolis is the capital of Mexico and one of the largest, and most polluted, urban centers in the world. The city sits at a high elevation in a mountain-ringed

Incineration of garbage contributes to Mexico City's blanket of pollution. Jorge Uzon/AFP/ Getty Images.

basin, which traps heavy pollutants in the atmosphere and helps to create a permanent, thick layer of brown smog.

At one time, scientists believed that the smog came from Mexico City's huge fleet of taxis, buses, motorbikes, and private cars, most of which had no emission-control equipment. By using x-ray and spectrometric analyses on particles captured from the air, the experimenters discovered what chemicals were being emitted, what time of day these pollutants appeared in heavy concentrations, and what effect the wind and weather had on the chemical haze that blanketed the city. One experiment revealed an unexpected pollution culprit: garbage incinerators and toxic lead.

The measuring equipment showed high concentrations of lead that increased during rush-hour mornings and remained low during holiday weekends. "We wondered what the source of the lead was," reports researcher Ryan Moffett. "Lead has been completely banned in gasoline sold in the city since 1997, which meant the aerosols were coming from something else." Also, the lead concentration correlated with high concentrations of chlorine and zinc. Waste incinerators, as the researchers knew, yield these toxic elements as well as lead, mercury, and phosphorus.

Incinerators in the northern neighborhoods of Mexico City are emitting these chemicals into the atmosphere, and winds from the north are spreading them throughout the city. The electronic equipment being burned in these incinerators is also contributing to toxic pollution. Instead of recycling cell phones, computer equipment, and the like, people are simply throwing them into the garbage to be burned, a process that releases chlorine as well as mercury and other poisonous heavy metals into the atmosphere. Moffett concludes: "Incineration—especially of discarded electronics, which are loaded with heavy metals and chlorine—is a dangerous process and a growing problem in developing countries."[6]

Case Study: Semiconductor Plants in California

Semiconductor manufacturing is big business in California. Eighty-five factories in Silicon Valley and other regions build processor, memory, and logic chips for computers; cell phones; and other electronic devices. But these factories put out a wide range of greenhouse gases. Their manufacturing processes also use fluorinated gases, which trap heat in the atmosphere at a far higher rate than carbon dioxide.

California passed its own global-warming law, known as Assembly Bill 32, in 2006. Governor Arnold Schwarzenegger supported and promoted the bill, claiming that California was leading the way in the combat against global warming and setting an example for other states to follow. The law set a goal of reducing greenhouse gases to 1990 levels by the year 2020.

Factory managers and company presidents protested the new mandate, which they claimed would cost their businesses almost

$40 million. The law, they argued, while offering little or no economic benefit to the state, would cost jobs if it forced factories to shut down. John Dale Dunn, a policy advisor for the Heartland Institute, commented that California was "chasing after a phantom menace . . . controlling carbon dioxide emissions is very expensive, and if plants are forced to comply, some will say, 'We can't afford to do business in California.'"[7]

Semiconductor manufacturers believe that California targeted their industry because regulating large utilities, car manufacturers, and heavy industries such as cement making is considerably more difficult. "We have not added a new semiconductor chip-making facility in at least the past 15 years," said John Greenagel, a spokesman for the industry. "We're already not competitive, but this just adds to the burden."[8]

Case Study: Carbon Dioxide Production by the Cement Industry

The cement industry represents a barometer of a nation's economy. Cement goes into nearly all building and road construction; the more active an economy and the faster that economy grows, the more cement the country needs to produce. Fast-growing China, for example, produces about 45 percent of all new cement in the world. The production of cement is also a major contributor to the earth's concentration of greenhouse gases.

The raw material of cement varies from one plant to the next, but usually it includes silicon, iron oxide, limestone, and calcium. The minerals are extracted from stone quarries, crushed into an aggregate mixture, and then sent to cement plants for processing. At the plant, the aggregate is further crushed into a powder (dry) or slurry (wet) form and then fed into a huge kiln, where the mixture is heated to 2,735°F (1,500°C) to produce clinker, a material made of small, round nodules. During this process, limestone is converted to calcium oxide and carbon dioxide. Most cement-production kilns are heated by burning coal; others use natural gas or fuel oil. Waste tires, plastics, and sewage can also be transformed into fuels for production kilns.

After the clinker cools, heavy machinery grinds the nodules into a fine powder. Various additives, such as gypsum and lime, are poured into the mixture during finish grinding in order to obtain different kinds of cement. The entire process consumes large amounts of electricity as well as fuel. According to the EPA report "CO_2 Emissions Profile of the U.S. Cement Industry," ce-

Cement Sustainability

The cement industry is taking measures to lessen its impact on the environment. The final goal in this drive is "zero landfill," meaning no waste disposal of concrete in any amount, and in any form.

The goal is an ambitious one, as the cement industry is vital to all nations and potential waste concrete is present in millions of buildings, roads, and other infrastructure. In addition, few nations press for the recycling of concrete, which is not by itself a toxic material. Nevertheless, the technology is becoming increasingly available and the rising demand for lowered carbon emissions is giving impetus to the effort. In addition, the construction industry now favors recycling for its saving of transportation and disposal costs for waste concrete.

Concrete can be broken down into its component parts to provide aggregate material for new roads or in the manufacture of new concrete. Newer equipment can crush concrete in any form, whether it is plain concrete or reinforced with steel rods. Recycling also uses much less energy than mining and processing of the raw materials used in concrete manufacturing.

Recycled concrete is growing in popularity and is being used in stadiums under construction for the London Olympic Games of 2012. The higher quality of recycled concrete—as well as recycled bricks, lumber, roofing, and other building components—is allowing designers of buildings a new source of materials as well as a favorable green public image.

ment making produces about 4 percent of all greenhouse gases emitted by the United States and 5 percent of all carbon dioxide emitted on the planet. In most developing countries, where heavy industry and utilities are not as vital to economic growth as the cement industry, the percentage of greenhouse gas emissions produced by cement making is even higher.

Green technology is available to reduce CO_2 emissions at cement plants, mostly from more efficient burning of fuel. Many outmoded plants in developing countries have been renovated by outside investors. But refitting cement plants also makes them more productive, and more production of cement means more greenhouse gas emissions from the basic chemical processes involved. The French company Lafarge, for example, has made new investments in eastern Europe, Russia, and China. According to the *New York Times* article "Cement Industry Is at Center of Climate Change Debate," the company has managed to reduce carbon dioxide emissions from 763 pounds (346 kg) per ton of cement produced in 1990 to 655 pounds (297 kg) in 2006. But Lafarge official Olivier Luneau also warned the *Times* that "Demand is growing so fast and continues to grow, and you can't cap that. Our core business is cement, so there is a limit to what we can change."[9]

European carbon-trading programs also help Lafarge purchase and refit cement production plants in eastern Europe. The company earns carbon credits for reducing greenhouse gas emissions and can then sell these credits on the carbon market. In many nations where cement production takes place, however, there is no limit to the amount of emissions a single plant can generate. As a result, production and emissions both rise in nations such as Ukraine, where demand for cement is high.

In the fastest growing nations such as China or India, economic growth (and cement production) takes priority over limiting greenhouse gas emissions. In these countries there is no political or economic incentive for refitting old cement plants. For that reason, international agreements such as the Kyoto Protocol (which went into effect in 2005, and which limits greenhouse gas emissions in developed nations) have little chance of success in lowering greenhouse gas emissions globally.

Notes

1. The California Report, "Japan: A Model of Energy Efficiency?" October 5, 2009. www.californiareport.org.
2. Julia Reinaud, "Climate Policy and Carbon Leakage: Impacts of the European Emissions Trading Scheme on Aluminum," International Energy Agency, October 2008.
3. International Energy Agency, "Tracking Industrial Energy Efficiency and CO_2 Emissions," 2007. www.iea.org.
4. Earth Observatory, "Global Fire Monitoring." http://earthobservatory.nasa.gov.
5. Hari Sud, "China: Industrialization Pollutes Its Countryside with Acid Rain," paper no. 1944, South Asia Analysis Group, September 9, 2006. www.southasiaanalysis.org.
6. "Tracking Down the Menace in Mexico City Smog," *Science Daily*, September 13, 2008. www.enn.com.
7. Celeste Altus, "California to Regulate Semiconductor Plants' CO_2 Emissions," The Heartland Institute, June 1, 2009. www.heartland.org.
8. Peter Fimrite, "State to Order Cuts in Greenhouse Gases," *San Francisco Chronicle*, February 27, 2009. http://capoliticalnews.com.
9. Elisabeth Rosenthal, "Cement Industry Is at Center of Climate Change Debate," *New York Times*, October 23, 2007. www.nytimes.com.

CHAPTER 3

The Future of Coal and the Nuclear Alternative

The mass production of electricity, which began in the early twentieth century, depended on generating plants that used water, and later coal, as a fuel source. By the end of the century, coal-fired plants were producing about 55 percent of all electricity generated in the United States. Coal is inexpensive and relatively easy to remove from the ground. Also, there is no scarcity of this fuel: Coal sources are expected to last several centuries even without new discoveries.

Coal-Fired Electricity-Generating Plants

Generating plants that use coal buy their fuel from large mining companies, which operate in the Appalachian region of West Virginia, Kentucky, and Pennsylvania, as well as in the Powder River Basin of Wyoming, which is now the largest source of coal in the United States. Coal is hauled to the plants via railroad, which is the only transportation method capable of handling the huge volume of coal needed by electric utilities.

At the plant, crushers grind the coal into a powder, which is then fed to a burner. The heat from combustion feeds water boilers, which generate steam. The steam drives turbines, which convert the heat energy to electrical power. The plant sends this power into the complex of transmission wires and transformers known as the electricity grid. In the 1940s, the cyclone fur-

Coal is transported up a conveyor belt into a coal-fired power plant near Fairfield, Texas. AP Images/David J. Phillip.

nace was adopted by many power plants, which could thereafter efficiently burn low-grade coal while producing less waste and pollutants. Cyclone furnaces reduced the amount of ash that escaped through plants' exhaust chimneys but also increased the amount of nitrogen oxides (NO_x) that were emitted.

Demand for electricity rose steadily through the middle of the twentieth century. The expansion of the population into suburban areas required the building of new electricity grids, and as manufacturing industries expanded, the need for power by factories also increased. Coal-fired plants burned hard and soft (brown) coal, emitting rising amounts of carbon dioxide (CO_2), sulfur dioxide (SO_2), nitrogen oxides and particulate matter (soot and dust) into the atmosphere.

The Creation of the Environmental Protection Agency and the Definition of an Air Pollutant

Although it was an important new law, the Clean Air Act passed in 1963 had no enforcement mechanism. Instead of establishing fines or penalties for polluters, it only authorized the Department of Health, Education, and Welfare to study the connection between air pollution and public health. If a company was found responsible for creating a threat to health with its toxic emissions, the government could sue the company. Otherwise, it could only offer advice and guidelines.

With amendments to the Clean Air Act in 1970, the government established the Environmental Protection Agency, or EPA. The law granted the EPA authority to define air pollution with a set of environmental standards for air, water, and soil. The EPA drew up National Ambient Air Quality Standards; the individual states then had to develop plans to reach those standards (the standards have been revised several times since then). The 1970 amendments also authorized the EPA to set emission limits for new vehicles and factories. The states were given the responsibility to monitor emissions from factories and power plants, and to enforce the EPA standards with fines and lawsuits.

Scientists and politicians debated the issue of acid rain and the regulation of the power industry throughout the 1980s. Individuals favoring a tax or regulatory mechanism argued that acid rain posed a problem for the environment and public health that the federal government should solve. Opponents believed that the problem of acid rain was exaggerated. They saw environmentalists as enemies of free enterprise and believed that more government regulation would hinder the development of new industry and technologies. In the view of opponents, a choice had to be made: strict regulation of emissions or a healthy economy.

Sulfur Dioxide Emission Allowances: The First Cap-and-Trade Program

In 1990, Congress passed more amendments to the Clean Air Act. Through this law, coal-fired power plants were granted emission allowances based on the amount of sulfur dioxide they were releasing into the atmosphere. Each allowance represented

THE ACID RAIN SUCCESS STORY

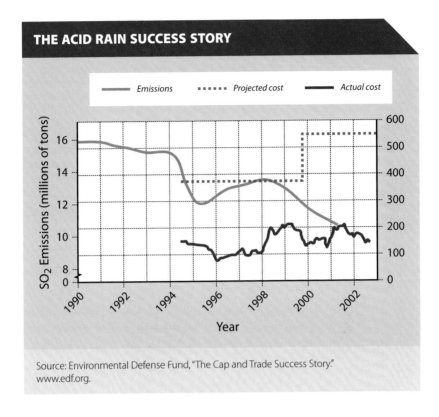

Source: Environmental Defense Fund, "The Cap and Trade Success Story." www.edf.org.

a single ton of sulfur dioxide emissions. If a plant reduced emissions below a certain level, the plant could save the extra allowances to offset emissions at another facility, or sell the allowances to another company. The price would be determined by the market, much like the price of company shares on a stock market. If emissions exceeded the limit, a plant would have to buy allowances; if a plant emitted more sulfur dioxide than it held allowances for, it paid a fine of $2,000 a ton.

The Acid Rain Program, as it came to be known, set a goal of reducing sulfur dioxide emissions to half of their 1980 levels (as well as nitrogen oxides to 27 percent of 1990 levels). A national cap on emissions was also established: 8.95 million tons of total sulfur dioxide emissions in 2000 and in each year beyond. Without a national cap, there would be no effective way to enforce the program's goals within an allowance-trading system. The government would have to constantly issue new allowances, which would lose their value as the supply continued to increase.[1]

The Acid Rain Program continues as of 2010, and coal-fired power plants are still trading allowances. The mechanism gives power plants a strong incentive to upgrade or replace old equipment to reduce their SO_2 emissions. Hundreds of plants have installed scrubbers that remove sulfur dioxide and other pollutants. Many have also switched to natural-gas-powered boilers or replaced high-sulfur coal with low-sulfur coal.

Congress updated the Clean Air Act with the Clean Air Interstate Rule in 2005. This law requires further reductions in SO_2 emissions in the eastern United States, beginning in 2010. In the meantime, similar regulatory programs have cut sulfur emissions in Europe; Great Britain had already achieved an 80 percent cut by the end of the 1980s, while a strict government emissions limit forced coal-fired power plants to install scrubbers or switch to natural gas as a fuel. Acid rain remains a serious problem in the developing world, however, particularly in China, where outdated technologies and a lack of regulation allow heavy sulfur dioxide emissions to continue.

Around the world, coal is used by thousands of power plants and in heavy industries such as steelmaking.

By any measure, the Acid Rain Program has been a success. Sulfur dioxide emissions have fallen, while the market price of SO_2 allowances has held steady between $100 and $200 per ton, well below early expectations. The program attained a compliance rate of 100 percent within ten years of its start; in fact, SO_2 emissions fell 22 percent (7.3 million tons) below the maximum levels mandated by Phase I of the program. The overall expense of the program, according to the federal Office of Management and Budget, has been between \$1.1 billion and \$1.8 billion a year—about 20 to 30 percent of the forecasted cost.[2]

The Acid Rain Program has successfully limited sulfur dioxide emissions, but coal remains an important energy source. It powers more than one third of all power plants in the United States; around the world, coal is used by thousands of power plants and in heavy industries such as steelmaking. According to scientist James Hansen, a prominent authority on global warming, the burning of coal has been responsible for about half of the increase in carbon dioxide emissions since the industrial revolution, and it is currently responsible for half the carbon dioxide emissions from anthropogenic (human or artificial) sources into the atmosphere.[3]

Clean Coal Technologies

Many older power plants are reaching the end of their useful lives. With the drive for reduced carbon emissions, a variety of clean coal methods have emerged in recent years that are designed to improve energy efficiency and lower greenhouse gas emissions by the power industry.

Clean coal is not a new idea that came in with the global warming issue. Electric plants have "washed" their coal for decades by processing it before burning, a step that reduces the emission of sulfur dioxide. Electrostatic precipitators are filtration devices—built into power plants' release mechanisms—which reduce the amount of ash escaping from plants through their smokestacks. Updated equipment also helps power plants to burn coal more efficiently, allowing the plants to generate more power for every ton of coal they use. In addition, recycling companies can use the waste products from coal burning—known as fly ash and bottom ash—to make a replacement for cement as a construction material.

Coal gasification is another form of clean coal technology. This process generates a fuel known as syngas from a process of steaming and oxidizing coal at the mining site. The chemical products of the process are hydrogen and carbon monoxide; while hydrogen can be used as fuel, carbon monoxide can be buried or sequestered underground in stable, deep fissures. Burning syngas results in much lower emissions of sulfur dioxide and nitrogen oxides than burning processed coal. In addition, some researchers believe that the hydrogen produced from coal gasification may eventually replace coal-burning as a large-scale power source.

Carbon Capture and Storage

Carbon sequestration belongs to a newer set of technologies generally known as carbon capture and storage, or CCS, which aims to remove the majority of carbon dioxide emissions from the coal-burning process. Capture refers to the removal of carbon dioxide before it reaches the atmosphere, whereas storage and sequestration are terms for burying the carbon in a stable form underground, where it has no effect on the earth's atmospheric CO_2 levels.

Engineers have developed several ways to separate carbon dioxide from other compounds. One of the more common methods is amine scrubbing. Like any other substance, coal gives off a stream of gas as it is burned. If passed through an amine solution, which is derived from ammonia, this gas gives up some of its carbon dioxide content before reaching the atmosphere. The solution can be reheated at a later point for a controlled release of the carbon dioxide.

Carbon capture by the amine solution method is not yet a practical solution to carbon emissions at power plants. The process is expensive and energy intensive; it captures only a part of the carbon dioxide generated; and the use of amine capture requires a great deal of energy by itself, both in the operation and in the manufacture of the needed components. In the interest of making this method more economical, the EPA is awarding several companies grants to develop the technology. The agency awarded $70,000, for example, to the IntelliMet company of

Future Power with FutureGen

FutureGen is an example of government working with industry to create sustainable energy projects on a large scale. The original plan for the FutureGen project was to design and build a coal-fired power plant that was nearly free of emissions. Once in operation the plant would produce hydrogen and up to 275 megawatts of electricity (enough to provide power to 150,000 homes) through coal gasification technology, while capturing approximately 90 percent of the carbon dioxide emissions for recycling or sequestration. It is a costly initiative, however, that ran into financial problems in 2008, when the federal government suspended support for the project. The Department of Energy, which was the federal agency responsible for public funding of the project, has since returned to FutureGen with new funding. Rather than building a new plant, the project will now retrofit an existing shuttered coal-fired Ameren power plant in Meredosia, Ill. for advanced oxy-combustion technology. The carbon emissions will be captured and conveyed through a pipeline to a storage facility in Illinois.

Missoula, Montana, to create a low-cost method of amine capture. According to an April 2010 EPA press release announcing the grant, "[This] is part of $2.38 million that will be awarded to 34 small businesses across the nation to develop innovative, sustainable technologies to help improve air quality, protect our water, work to decrease the effects of climate change, and support green jobs."[4] The federal government also provides tax breaks to help utility companies install and operate the equipment needed to reduce emissions of carbon dioxide and other greenhouse gases.

The gasification method employs high-pressure steam to burn coal, which produces hydrogen and carbon monoxide, which are then burned in a gas turbine to generate power. This method can be combined with the oxyfuel process, which changes the chemical environment of coal burning. Coal is burned in an atmosphere of pure oxygen, a process that creates a much higher concentration of carbon dioxide in the gas released, and that allows the amine-capture process to operate much more efficiently. In addition, the captured carbon dioxide can be sold to oil drillers, who can use the compound for enhanced oil recovery in fields where conventional drilling is no longer productive. Few new sources of crude oil have been developed in the last decade, and carbon dioxide recovery of hard-to-mine oil and natural gas fields has become a growing industry in the United States and around the world.

The oxyfuel process has not yet been widely adopted in the United States. Equipping an existing power plant with the necessary burners and turbines requires a large investment. Although the United States has an active market in sulfur dioxide emission allowances, carbon trading is still limited in comparison to Europe. Where carbon trading is active and has become an important factor in the operation of coal-fired power plants, there is more incentive to upgrade equipment.

Another important developing technology for reducing carbon emissions is carbon sequestration, in which carbon is captured from emission sources and buried. In geosequestration, carbon dioxide is injected into underground storage chambers created from mined-out fields of oil or natural gas, salt domes, or other natural or artificial formations. The carbon is sealed into the earth by layers of caprock, which is dense enough to prevent carbon or other gases from reaching the surface. There is no lack of suitable underground formations for the sequestration of carbon dioxide. Several thousand cubic miles of potential storage sites have already been discovered and mapped.

The process was first tried by a Norwegian gas company in the Sleipnir gas field underneath the North Sea, where about one million tons of compressed CO_2 is sequestered each year. The company made a large initial investment to develop the project but is also reducing the carbon taxes it pays to the government of

Norway, which has imposed such taxes on oil and gas producers since 1991. Around the world, other sequestration projects will make use of underground salt reservoirs or underground natural gas or oil fields that have already been mined.

Although a promising future technology, carbon sequestration still raises safety questions. The uncontrolled release of concentrated carbon dioxide is extremely dangerous; a natural eruption at Lake Nyos in Cameroon in 1986, for example, killed almost two thousand people. In addition, the injection of pressurized material into underground fissures and faults poses the danger of destabilizing underground geological structures, thus causing tremors and earthquakes.

Carbon may also, in theory, be sequestered by delivering it to the ocean floor via pipeline; the higher density of carbon dioxide at oceanic depths would keep it underneath the surface. Such oceanic storage, however, could affect marine life through the reaction of carbon dioxide with seawater, a process that would generate large amounts of damaging carbonic acid. The acidification of the ocean already concerns marine biologists, and oceanic storage may end up merely substituting one environmental problem for another.

Many environmentalists also oppose carbon capture, sequestration, and all other clean coal technologies on the grounds that they still require the mining of coal, a process that damages the environment and human health. Author James B. Martin-Schramm, in *Climate Justice: Ethics, Energy, and Public Policy,* cites a report that reveals that more than one hundred thousand coal miners have died in accidents, and more than twice that number from black lung disease, since 1900. In addition, strip mining of coal has destroyed as much as a million acres of hardwood forests in the eastern United States, and has destroyed nearly five hundred mountains in the last two decades alone.[5]

Instead of clean coal, utility companies, in this view, should be looking to alternative and renewable sources of energy, such as solar, wind, geothermal, and biomass. Many environmentalists also now support a power source that was once considered the most toxic and unsafe of all: nuclear power.

The Nuclear Alternative

Nuclear power was first developed in the 1950s. Many nuclear reactors draw energy from a controlled chain reaction of radioactive isotopes of uranium. The heat generates steam, which then operates electrical turbines linked to the power grid. Nuclear power is a clean technology that creates no toxic emissions under normal operation and delivers no greenhouse gases into the atmosphere. Nuclear power remains a nonrenewable resource, however, as the supply of useful uranium is limited. Nuclear energy also raises safety concerns over the possibility of an uncontrolled chain reaction that could release radiation into the atmosphere. Another fear is that stolen nuclear fuel stocks or waste could be used by enemy governments or terrorists to manufacture a nuclear weapon.

The major problem in nuclear power, however, is not emissions or theft but storage. Spent nuclear fuel rods remain radioactive for one hundred thousand years or longer and must be stored in a location that is geologically stable and far from populated areas. The technology for nuclear waste storage is still developing; in the meantime, the building of nuclear plants has become a tangled, complex process of permitting and review by public agencies, making it so expensive and difficult that no new nuclear plants have been built in the United States in more than thirty years. For this reason, the energy produced by nuclear power remains expensive as well, and the cost of electricity is comparable to that created by new coal or natural gas plants.

Outside the United States, the nuclear industry is still thriving. According to statistics published by the Nuclear Energy Institute, 28 countries are using nuclear energy to generate power, and 441 nuclear plants are in operation. Lithuania and France represent nuclear-energy leaders with more than 75 percent of electricity in those countries nuclear-generated. In late

New Advocates for Nuclear Power

Although environmentalists have campaigned as a group against nuclear power for decades, and have been successful in stopping construction of new nuclear plants, many individuals have changed their minds in the face of what they view a much greater danger: greenhouse gas emissions and global warming. In "Nukes are Green," a widely noted opinion piece published April 9, 2005, in the New York Times, *journalist Nicholas Kristof lays out the case for the nuclear alternative to fossil-fuel use.*

It's increasingly clear that the biggest environmental threat we face is actually global warming, and that leads to a corollary: nuclear energy is green.

Global energy demand will rise 60 percent over the next 25 years, according to the International Energy Agency, and nuclear power is the cleanest and best bet to fill that gap.

Solar power is a disappointment, still accounting for only about one-fifth of 1 percent of the nation's electricity and costing about five times as much as other sources. Wind is promising, for its costs have fallen 80 percent, but it suffers from one big problem: wind doesn't blow all the time. It's difficult to rely upon a source that comes and goes.

In contrast, nuclear energy already makes up 20 percent of America's power, not to mention 75 percent of France's . . . for now, nuclear power is the only source that doesn't contribute to global warming and that can quickly become a mainstay of the grid.

2010, 60 new plants were under construction around the world, with China alone accounting for 23. Perceiving nuclear as the best alternative to fossil-fuel plants, these nations have streamlined the permitting process and thus allowed new construction to proceed, even as safe and economically feasible storage alternatives are still being developed.

The 2005 Clean Air Interstate Rule: Further Emission Reductions

The EPA has continued to pass new regulations to control the emission of toxic pollutants into the atmosphere. In 2005, the agency passed the Clean Air Interstate Rule, or CAIR. It required twenty-eight states and Washington, D.C., to reach specific emission targets for sulfur dioxide and nitrogen oxides released by all of their electrical generating plants. The states had two options: require the plants to take part in a cap-and-trade system, or meet their emissions budgets through regulation enforced by taxes and fines, or by any other method of their choosing.

In addition, a new cap was placed on mercury emissions from coal-fired power-generating plants. Mercury is a highly toxic element that causes serious damage to the environment when released into the air and water, and it can pose a health hazard to humans who consume fish and other foods with high mercury levels. Although mercury has been banned from most consumer products and manufacturing processes, the burning of coal to generate electricity remains an important source of mercury emissions.

This Clean Air Mercury Rule aims to reduce mercury emissions from power plants, from 48 tons every year to 15 tons. The EPA patterned this new rule on its success with sulfur dioxide emissions and established a cap-and-trade program that would be implemented in two parts. Phase I, which runs from 2010 until 2017, will reduce emissions nationwide to 38 tons; Phase II limits emissions to the final goal of 15 tons. Each state covered by the rule has been assigned a proportional cap; Texas, for example, is now permitted to emit 4.65 tons of mercury of year until 2017, and thereafter 1.84 tons a year.[6] New coal-fired plants built after January 30, 2004, must meet stricter emissions controls than older plants.[7]

A national mercury-emission allowance system is now in place. Each state has an assigned mercury budget and must submit a plan to the EPA that describes how the state intends to reduce mercury emissions from coal-fired power plants. If it succeeds, the Clean Air Mercury Rule will give a strong impetus to the passage of a nationwide cap-and-trade system for greenhouse gas emissions.

Notes

1. Environmental Protection Agency, "History of the Clean Air Act," November 16, 2010. www.epa.gov.
2. Environmental Defense Fund, "The Cap and Trade Success Story," February 12, 2007. www.edf.org.
3. James E. Hansen, "Testimony Before the Iowa Utilities Board," October 2007. www.columbia.edu.
4. Environmental Protection Agency, "EPA Awards $140,000 for Technology Development to Small Businesses in Bozeman, Missoula," April 5, 2010. http://yosemite.epa.gov.
5. James B. Martin-Schramm, *Climate Justice: Ethics, Energy, and Public Policy*, Minneapolis: Augsburg Fortress, 2010, p. 7.
6. Texas Commission on Environmental Quality, "Clean Air Interstate Rule and Clean Air Mercury Rule," April 9, 2010. www.tceq.state.tx.us.
7. Environmental Protection Agency, "Clean Air Mercury Rule: Basic Information," January 20, 2010. www.epa.gov.

CHAPTER 4

Transportation Industries

Transportation industries are vital to the economic well-being of any nation. Without the ability to transport people and goods via road, rail, and air networks, most economic activity would stop. As a result, regulation of the transportation sector becomes a subject of intense public debate and legislative wrangling. To private citizens, especially in the United States, the automobile is the key to freedom and independence. For most businesses, the cost of moving goods to market is a significant expense, and any increase in that expense can mean a loss of income and profit.

Yet private and public transportation sources, including passenger cars and trucks, buses, trains, aircraft, and watercraft, are emitting about 30 percent of all greenhouse gases produced in the United States, and they account for half the growth in total greenhouse gas emissions since 1990.[1] Any comprehensive solution to the issue of global warming includes regulation of the transportation industry in some form: carbon taxes, a cap-and-trade system, or stricter fuel-economy standards.

Vehicles Contribute to Rising Greenhouse Gas Concentrations

For more than a century, the automobile industry has been selling vehicles that allow a sense of mobility and ease that would have been unthinkable just a few generations before. In the United States, a car culture emerged in the early twentieth century as automobiles were used for commuting, for leisure travel, or to visit friends and family in distant places. After a house, the family car became the most important status symbol, a sign of wealth and success. This calculation remains true today in emerging market economies such as India and China, where the principal economic and social dividing line now is the one between persons who own a private car and those who do not.

Private vehicles may be expensive, but they have become absolutely necessary for millions of people who could not get to work, or many other places, without them. During the last fifty years, most new housing in the United States has been built with the automobile in mind. Suburbs have grown in areas distant from inner cities and inaccessible to mass-transit systems. Streetcars have been retired, and passenger trains have grown obsolete; intercity buses have dwindled in number and, in most cities, people do not use taxicabs. The passenger vehicle endures, with only air travel serving as a popular alternative for long-distance trips.

To avoid reliance on imported oil, the energy and automobile industries have been working on alternative fuels for many years. These two sectors must work cooperatively, as most alternative fuels require different kinds of engines.

But cars also release toxic pollutants into the atmosphere, as well as greenhouse gas emissions that contribute to global warming. The use of gasoline refined from crude oil also creates economic and political problems for nations that do not have domestic sources of oil. In an attempt to find a solution to global warming and oil dependence, car manufacturers are creating new kinds of vehicles that use new kinds of fuel.

Making Cars More Efficient: Alternative Fuels

Most car engines run on gasoline, a fuel derived from crude oil. Gasoline is produced at refineries, transported via large highway tankers, and sold at service stations. The internal combustion engine used by automobiles produces carbon dioxide (CO_2), sulfur dioxide (SO_2), nitrogen oxides (NO_x), and particulate matter, which contribute to smog and haze over large cities.

In most countries, the production of gasoline depends on imported crude oil, which creates an economic and political hazard in countries that don't have their own oil sources. To avoid reliance on imported oil, the energy and automobile industries have been working on alternative fuels for many years. These two sectors must work cooperatively, as most alternative fuels require different kinds of engines. One exception is ethanol, a fuel that, in theory, can be produced abundantly in the United States and elsewhere and that can be burned in any conventional internal combustion engine.

Ethanol and Other Biofuels

Ethanol was one of the earliest alternative fuels developed. In some countries, such as Brazil, it has become a major source of fuel for private and public vehicles. It is distilled from fermented starches contained in certain vegetables crops, such as corn, or from certain plants and trees. The use of ethanol helps reduce dependence on imported crude oil and results in lower greenhouse gas emissions. In addition, ethanol is relatively inexpensive in areas such as the US Midwest, where corn and other useful crops are abundant, and more expensive in areas where it has to be transported a greater distance to be sold.

Ethanol in its pure form cannot be used in most automobile engines, and for commercial sale must be blended with traditional gasoline. A common form is E10 ethanol, which is sold in the United States and is a mixture of 10 percent ethanol and 90 percent gasoline. Although traditional internal combustion engines can use E10, the mileage is usually not as high as with pure gasoline.

E85 is another form of ethanol. New flexible-fuel vehicles (also known as FFVs) can use E85, gasoline, or any combination of these two fuels. E85 has a lower energy content than regular gasoline, however, and results in poorer mileage, whether used in a traditional car or an FFV. Car owners also have to search out gas stations supplying E85, which has limited availability. In many areas, the drive needed to obtain E85 for a car or truck negates any cost savings and further raises the amount of greenhouse gas emissions created by the vehicle.

Ethanol belongs to the family of biofuels, which are fuel sources derived from organic matter such as plants, trees, grasses, and crops. Biodiesel—made from oil, grease, fat, or common

This production plant in South Dakota refines corn into ethanol—one of several biofuels currently in use. © Jim Parkin/Alamy.

crops such as soybeans or flax—is the most important biofuel in Europe and is used to reduce the carbon dioxide emissions from diesel fossil fuel. Diesel-powered engines, which burn fuel more efficiently than engines using conventional gasoline, have long been in widespread use in the nations of Europe, most of which have few domestic sources of oil. In the United States, diesel engines are used by most buses and trucks, whereas most cars still run on oil-derived gasoline.

Natural Gas and Propane

Formerly burned off as a useless by-product of oil drilling, natural gas now has an international market as a fuel for machinery, power plants, and home heating systems. In recent years, compressed natural gas and liquefied natural gas have become important alternative fuels for vehicles. Another cleaner-burning fuel is liquefied petroleum gas, also known as propane or LPG. Bifuel cars and trucks can use either natural gas/LPG or gasoline. Although there are very few passenger cars made for natural gas or propane, it is possible to retrofit cars and trucks to make their engines natural-gas compatible.

Natural gas and LPG are less expensive, mile for mile, than gasoline. They also burn cleaner, producing less smog and greenhouse gas emissions. The United States has plentiful supplies of natural gas and LPG and would have little need to import them from foreign countries even if demand grew in the years to come. There are few mass-produced natural gas or LPG cars, however, and these alternative fuels pose a logistical problem for owners: Most service stations don't provide them. In addition, bifuel vehicles need to be fitted with separate fuel tanks, which reduce the amount of room available for passengers and cargo.

The Uncertain Future of Hydrogen

The future of vehicle fuel may be in hydrogen, one of the most abundant elements on the earth and one that burns without emitting greenhouse gases. There are many domestic sources for hydrogen fuel, and in theory hydrogen would not have to be imported from any foreign country. Although hydrogen fuel has been used for rocket propulsion since the 1950s, its use in wheeled passenger vehicles remains in the early, theoretical stages.

To provide energy, hydrogen must be stored in a compressed, liquid form, which it can attain only at temperatures colder than $-420°F$ ($-215°C$). When the hydrogen is allowed to warm up, it transforms into a gas, which is then piped to a fuel cell where it is ionized and its freed electrons are channeled through a wire, producing an electric current that can do work, such as turn an electric motor.

Hydrogen fuel is subject to the same rules of supply, demand, and market share as other fuels. It is less efficient by weight than gasoline or natural gas, and cars using hydrogen have a limited range between fill-ups. The production of hydrogen fuel for passenger cars is still expensive, and very few sales outlets exist. Cars using hydrogen fuel cells remain unaffordable for most people, and until a wider market exists for this fuel, automakers will not mass-produce hydrogen-burning cars: the expenses of production will not be met by the income from sales.

Hybrid Cars

New types of passenger cars are now in production as the market develops new fuel sources. In order to gain acceptance in the marketplace, however, these cars have to meet some minimum specifications. In general, the automobile market demands that cars should be able to travel at least 300 miles before refueling or recharging. They must maintain average speeds on the freeway system of about 70 miles per hour in uncongested and rural areas. They should also be easy to refuel, and there should be plentiful sources of needed fuel or electrical charge.

The gasoline-electric hybrid is the first alternative-fuel vehicle to meet these specifications. The hybrid, which combines gas-powered and electrical engines, overcomes the most serious drawbacks of electric vehicles: their limited range and the time and trouble of recharging them. Hybrids have electric motors that serve both to power the vehicle and to store electrical charge. The motors draw on charge when accelerating and return charge to the storage batteries when slowing down.

By 2010, the hybrid car made up about 2.5 percent of the US car market and was slowly gaining acceptance among the car-buying public. According to industry analyst Alan Baum, the US car market offered more than 50 hybrid car models by 2010, as well as pure electric-drive vehicles such as the Chevy Volt. With a forecast for 108 hybrid and electric-drive vehicles to compete in the market by 2015, Baum stated, "In the past few years, major automakers have changed their strategy. They used to fight fuel efficiency regulations. Now, as the numbers go up and time marches on, automakers are saying, 'Look at the list of models. We're doing our part. We're bringing a raft of new vehicles and technologies.'"[2] Stricter fuel-economy standards and the rising cost of gasoline may make hybrids and plug-in electric cars practical, economic, and popular.

The Debate over CAFE Standards

The automobile industry, like other manufacturers and utilities, is subject to environmental regulation. The most important control on new cars is the set of fuel efficiency limits known as corporate average fuel economy, or CAFE. These standards apply to cars, trucks, vans, and sport utility vehicles. They originated in 1975, shortly after the Organization of Petroleum Exporting Countries, a group of oil-producing nations, placed an embargo (ban) on oil exports to the United States and other countries. The embargo arose from conflicts in the Middle East; although it eventually ended, it raised concern that the United States could be held as an economic hostage by nations that control a vital source of oil.

The CAFE standards were followed by a zero-emission vehicle (ZEV) program enacted by the California Air Resources Board in 1990. The ZEV program granted credits to automak-

New EVs on the Market

The US auto industry, which nearly collapsed in the recession that began in 2008, is reviving with plans to build low-emission and zero-emission vehicles. Hybrids and electric cars are now widely available to buyers; Ford is preparing to mass-produce the Focus EV (electric vehicle) in 2011, but the industry has competition from Europe and Asia. Nissan has already brought the electric Leaf to market. This small EV hatchback, which became widely available in 2010, runs on electricity and has a range between charges of about 100 miles. Mercedes-Benz, Volkswagen, BMW, and Renault all have EVs on the drawing board and in limited production runs. In China, the BYD ("Build Your Dreams") E6, manufactured by a battery company that began designing cars in 2003, has a range of up to 250 miles and can reach a top speed of 100 miles per hour. The car has not yet reached the US market, however.

The EV still has its drawbacks for the average car buyer. The cars are relatively expensive, and they require recharging at home after reaching the limit of their short range (a network of public recharging stations will take many years to implement). These limitations will continue to be troublesome in the United States, where many passenger cars are regularly used for long-distance trips.

ers for cars whose operation created no toxic pollution or greenhouse gases.

The national fuel economy standards, as well as limits on the emission of carbon dioxide and other greenhouse gases, are set by the EPA. Fuel economy is measured in minimum miles per gallon attained by the entire fleet of cars and trucks produced by a single automaker. If the manufacturer fails to meet the standards, it must pay a fine of $5.50 for every tenth of a mile per gallon under the minimum, multiplied by the number of vehicles the automaker manufactures for domestic sales.

The CAFE standards have always been the subject of controversy between the automobile industry and government. During the 1970s and 1980s, many manufacturers complained that CAFE standards were forcing them to make smaller cars and trucks, which, at that time, had limited appeal among consumers. In the meantime, California set stricter standards of its own that posed a dilemma to automakers, who had to build their cars to a different specification for sale in that state.

In 2010, new CAFE standards set the bar even higher, placing a minimum fuel economy of 35.5 miles per gallon on new vehicles, a figure to be attained by the year 2016. The EPA announced that this new standard would cut greenhouse gas emissions by 30 percent.[3] Automakers making electric vehicles can claim partial credit against their fossil-fuel fleet in meeting the standards. Small carmakers, such as Porsche and Jaguar, will not be subject to the new standards until 2017, but in the meantime they must buy credits from larger manufacturers who exceed the fuel emission caps set down by the EPA.

The fuel-economy standards were based on ones already passed in California in 2004. They are being met by automakers that are designing cars with lighter materials, smaller and more efficient engines, and transmission systems with as many as eight speeds, rather than the standard three or four. The new national standards going into effect in 2016 will solve this problem—unless California passes new standards in 2017—and are generally supported by the auto industry. "This is an example of where the federal government has actually done something right," one spokeswoman for the Alliance of Automobile Manufacturers stated. "A year ago, we were facing piecemeal policies set out by EPA, DOT [Department of Transportation], and groups of different states. Our auto engineers cannot design vehicles to different standards."[4]

Public Transportation: The Hybrid Bus and Light Rail

Many cities around the world have adopted low-emission vehicles or zero-emission vehicles for their public transportation fleets.

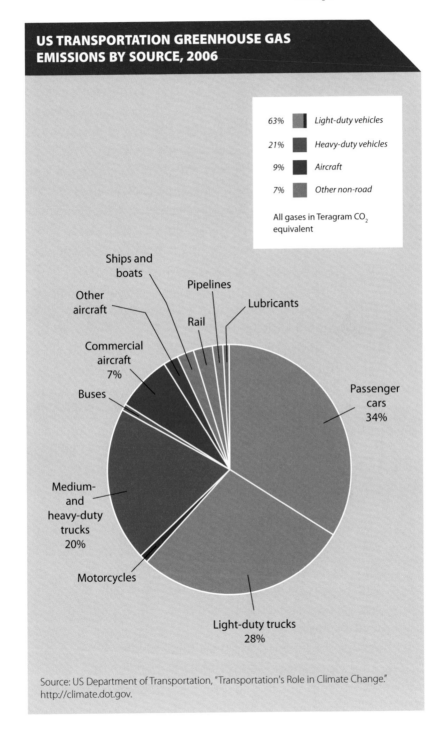

US TRANSPORTATION GREENHOUSE GAS EMISSIONS BY SOURCE, 2006

63% Light-duty vehicles

21% Heavy-duty vehicles

9% Aircraft

7% Other non-road

All gases in Teragram CO_2 equivalent

Ships and boats

Pipelines

Lubricants

Other aircraft

Rail

Commercial aircraft 7%

Passenger cars 34%

Buses

Medium- and heavy-duty trucks 20%

Motorcycles

Light-duty trucks 28%

Source: US Department of Transportation, "Transportation's Role in Climate Change." http://climate.dot.gov.

Fast intercity electric trains have been running for many years in Europe and Japan, while light-rail mass transit is a common feature of large cities in Denmark, Germany, and the Netherlands.

The United States is slowly catching up. California set strict clean-air regulations that barred the use of diesel-burning buses, and as a result California cities have been forced to acquire hybrids. The city of New York has replaced about two thousand of its conventional fossil-fuel buses with hybrids, equipped with series-drive engines that draw on diesel power as well as electric batteries. Hybrid buses, like most low-emission vehicles, require a large initial investment. They cost about $550,000, versus about $400,000 for a traditional bus.[5] But the hybrid buses are quieter, and their operation reduces new emissions.

Light-rail systems are also gaining in popularity and are meeting the demand for public transport in denser areas of cities such as Minneapolis, Salt Lake City, and Seattle. Light-rail systems have come under fire, however, for their high initial cost and the fact that they do not save on collective energy use by a city's residents. Light-rail runs on electricity, which must be generated by fossil-fuel-burning power plants. One study carried out by the National Transit Database measured Salt Lake City's transport energy use before and after light rail, using British thermal units (BTUs). The study found that the old system used 4,300 BTUs for every passenger mile, whereas the new combined light-rail/bus system uses 5,574 BTUs.[6]

The result is higher use of energy and burning of fossil fuels to create that energy. Of course, if commuters are using light rail instead of passenger cars, an energy savings takes place. But in Salt Lake City and other large metropolitan areas, light-rail systems have made a minimal impact on commuting habits. Instead of building expensive light-rail systems to replace busy bus routes, proponents of traditional public transportation believe that cities should simply extend their bus networks to service more commuters and provide more frequent service. "Considering the initial $1.4 billion capital cost of the [light-rail] system, Phoenix has . . . spent over $70,000 per daily rider just to build the system," commented blogger Warren Meyer on Forbes .com in September 2010.

> This is an astonishing number—no wonder the riders of the system love it! The taxpayers of Arizona bought rail riders a commuting vehicle that costs nearly three times the $23,800 list price of a Prius III hybrid. The city could have, rather than build the system, bought every regular daily rider a new car and still had nearly a billion dollars left over—and those who got the car would have had a transportation option that went anywhere in the city, not just to 28 stops along a single 20-mile line.[7]

Aircraft Emissions

Cars and buses are not the only public vehicles emitting greenhouse gases. The airline industry shares some responsibility for global warming but has largely escaped any form of regulation or control of its emissions. In the next environmental treaty after the Kyoto Protocol, an international agreement to limit carbon dioxide and other greenhouse gas emissions, that status may change.

The airline industry has been growing steadily since the 1990s. Low-budget airlines allow people to fly between many cities at a cost much lower than that of driving or taking passenger trains. The number of commercial aircraft has doubled in the last twenty years and is expected to continue growing. Aircraft emissions still make up a small part of total greenhouse gas emissions, estimated at about 3 percent, within the EU;[8] in the United States, aircraft contribute 12 percent of emissions from all transportation sources.[9] Emission levels are growing faster among aircraft than in any other transportation or industry sector.

There are no limits or trading mechanisms for aircraft emissions, which were not included in the Kyoto Protocol. But airline companies, like car manufacturers, want to enjoy a reputation as being environmentally friendly. Several European companies are joining the carbon-trading mechanisms already in place in Europe. They are expected to try to offset carbon dioxide emissions with investments in new, low-emission technology as well as alternative fuels (all aircraft currently run on aviation fuels refined from crude oil). New aircraft designs will feature planes that weigh less and have improved aerodynamics, which will reduce fuel consumption. The new Boeing 787, for example, is made of composites that make it more efficient than older aircraft. On the ground, air traffic management can also be revamped to reduce the amount of time airplanes taxi from gates to runways. Airlines can also consolidate or change their routes in order to reduce the amount of time and distance between destinations.

Ship Emissions

Commercial shipping, vital to international trade, is still largely free of emission controls. Large cargo ships use diesel-powered engines and emit pollutants in busy ports such as New Orleans, Tampa, New York, and Long Beach, California. But shipping companies have strongly protested restrictions on their emissions, especially by individual states. If the states or port cities were to begin setting different emissions standards, so the argument goes, shipping companies would have to change their operations or their equipment every time they move from one harbor to the next.

The answer is an international treaty effort that began in 1972 with the adoption of the Convention on the Prevention of Marine Pollution by Dumping of Wastes and Other Matter, also known as the London Convention. This treaty, which now has eighty-one nations participating, restricts the dumping of waste matter by ships and other oceangoing structures such as oil-drilling rigs, as well as by aircraft. In 1993, it was extended to ban the dumping of radioactive waste.

Greenhouse gas emissions from ships, however, still have not come under the control of any international agreement. This problem remains in ports all over the world. Long Beach, together with the port of nearby Los Angeles, for example, admits about 40 percent of all goods imported into the United States. Foreign shipments through this area create greenhouse gas emissions not just from ships but also from the trucks that receive the goods for transport via the interstate highway system.

To reduce the environmental impact of oceangoing ships, California has passed rules that require diesel-powered ships to change to low-sulfur fuel (with a sulfur content of less than 15,000 parts per million) whenever they are sailing within 24 nautical miles of the coast (a nautical mile equals about 1.15 statute miles or 1.85 km, the distance measurements used on land). This cleaner fuel reduces the release of sulfur dioxide, nitrogen oxides, and particulate matter over the basin that includes Los Angeles and its many satellite cities. California is also planning further restrictions on fuel use by oceangoing ships in 2012, when the sulfur content must fall below 1,000 parts per million whenever the ship sails within 200 miles of shore. The EPA is planning to match the California standards and require that large ships adopt updated pollution-control technology, beginning in 2015.[10]

Notes

1. Environmental Protection Agency, "Transportation and Climate," September 14, 2010. www.epa.gov.
2. "Expert: Expect More Than 100 Hybrid and EV Models in U.S. by 2015," HybridCars .com, September 20, 2010. www.hybridcars.com.
3. John M. Broder, "U.S. Issues Limits on Greenhouse Gas Emissions from Cars," *New York Times*, April 1, 2010. www.nytimes.com.
4. Broder, "U.S. Issues Limits."
5. Bill Sweet, "New York Leads World in Hybrid Bus Adoption," *IEEE Spectrum*, April 14, 2010. http://spectrum.ieee.org.
6. Randall O'Toole, "Reducing Greenhouse Gas Emissions," The Antiplanner, April 18, 2007. http://ti.org.
7. Warren Meyer, "Urban Light Rail Fail," Forbes.com, September 22, 2010. http://blogs .forbes.com.
8. Eilene Zimmerman, "Aviation's Greenhouse Gas Emissions Could Triple in the Next 50 Years," True/Slant, June 1, 2010. http://trueslant.com.
9. Center for Biological Diversity, "Aircraft Emissions." www.biologicaldiversity.org.
10. Amy Littlefield, "U.S., California Programs to Reduce Ship Emissions," *Los Angeles Times*, July 2, 2009. http://articles.latimes.com.

CHAPTER 5

The Electrical Power Grid

Although natural philosophers have studied electric pheno-
mena since ancient times, there was no way to generate
and transmit electrical energy over large distances until the
nineteenth century. The first commercial generators derived
their power from the movement of water (hydropower), an idea
adapted from the traditional water mill. Engineers created a
method of capturing that electricity and sending it to distant locations via transmission lines. These lines linked electrical plants
to homes, offices, and factories, providing these end-users with a
reliable source of power.

Hydropower and Coal

Hydropower was not available to everyone; areas without a source
of running water were, for a time, without electricity. Eventually,
generators were developed to run on fossil fuels such as coal, oil,
and natural gas. When the fuel is burned within a water boiler,
the steam that results drives turbine blades, which generate electrical energy. Nuclear reactors work on the same principle, but
their power source is the energy released by subatomic reactions.
A majority of the electrical power produced in the world is now
created by steam turbines.

Commercial electricity brought about a second wave of development in the industrial revolution. Electricity was a cleaner
and safer source of light and heat than coal or oil; it was cheaper

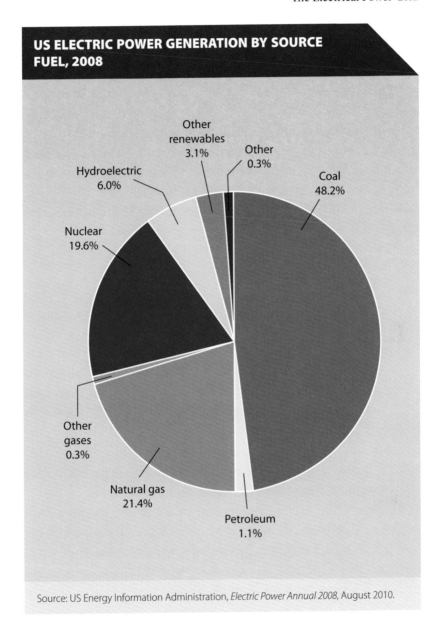

US ELECTRIC POWER GENERATION BY SOURCE FUEL, 2008

Other renewables 3.1%

Other 0.3%

Hydroelectric 6.0%

Coal 48.2%

Nuclear 19.6%

Other gases 0.3%

Natural gas 21.4%

Petroleum 1.1%

Source: US Energy Information Administration, *Electric Power Annual 2008*, August 2010.

and, once an electrical grid was built, widely available to people in the city or the country. For these reasons, demand for electricity rose steadily through the twentieth century. Large appliances, air conditioning systems, televisions, and computers all created a greater need for electricity among consumers, and a reliable supply of electricity became absolutely essential to developing societies. Without electric power, most factories and offices couldn't operate, and families would be hard-pressed to survive in private homes. Electricity powers fuel pumps used by service stations, ATMs needed for bank transactions, and communication systems needed for phone calls and messages. A lengthy interruption in electricity would bring a modern, technologically advanced society such as the United States to a standstill.

Demand for electricity rose steadily through the twentieth century . . . and a reliable supply of electricity became absolutely essential to developing societies.

Yet storing large amounts of electrical energy is still costly. Electric utilities must match demand with supply, and they must use a network of transmission lines that can draw on reserve or backup systems when electricity demand rises. In the past, the expense of storage and the need for large backup systems made "green" generation of electricity from renewable sources, such as wind and solar, impractical for large utility companies.

Whereas the United States and other countries have built large electrical grids, developing nations such as India and China are still in the process of fully electrifying rural areas and building enough generating-capacity—using fossil fuels and nuclear power—to meet demand. As these nations continue to grow at a rapid pace, their need for electricity rises as well. New apartments require lighting, heat, and air conditioning; new factories consume large amounts of power to drive assembly lines, robotic systems, and office equipment. Thus, nations such as China see any restriction on fossil-fuel use as seriously harmful to their industries and economies.

The Advent of Nuclear Power

With the engineering of atomic weapons during World War II, scientists realized they had unlocked an extremely powerful form of energy. After the war, civilian nuclear power projects were developed in the United States, Japan, Europe, and the Soviet Union. Nuclear power held out the promise of cheap, clean, potentially limitless amounts of electrical energy. Despite hazards to public health and environmental dangers from radioactivity, nuclear energy also qualified as a clean energy source, because nuclear reactions do not release carbon dioxide or any other greenhouse gases into the atmosphere.

The nuclear power process is based on the transformation of mass within the atomic nucleus into energy. The nucleus is composed of protons and neutrons. When the nucleus is bombarded with neutrons, it divides into separate and lighter nuclei. This event, called fission (meaning splitting), releases large amounts of energy. The most important fuel for nuclear reactors is the element uranium-235 (an isotope of uranium that has 92 protons and 143 neutrons in the nucleus), which has a potential energy content about 3 million times that of the equivalent weight of coal.

Nuclear power plants use the heat generated by nuclear reactions to power steam-driven turbines in a process similar to coal-fired or oil-fired generating plants. Nuclear reactors produce radioactive waste, however, which must be stored in safe locations, as the atoms of radioactive waste remain toxic for a very long time. If these atoms are released into the atmosphere, they can cause cancer and genetic mutations, and in high doses severe illness and death.

The problem of storage and the danger of radiation released by accidents at nuclear power plants have stalled the nuclear power industry in the United States, where there have been no new nuclear plants completed since the Watts Bar 1 station went online in Tennessee in 1996. However, the demand for alternative energy has renewed interest in breeder reactors. Breeder

SCHEMATIC OF A NUCLEAR POWER PLANT

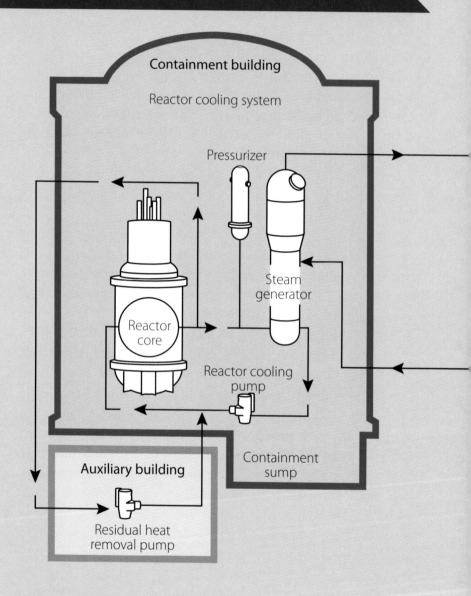

Containment building

Reactor cooling system

Pressurizer

Steam generator

Reactor core

Reactor cooling pump

Containment sump

Auxiliary building

Residual heat removal pump

There are two major systems utilized to convert the heat generated in the fuel into electrical power for industrial and residential use. The primary system transfers the heat from the fuel to the steam generator, where the secondary system begins. The steam formed in the steam generator is transferred by the secondary system to the main turbine generator, where it is converted into electricity. After passing through the low pressure turbine, the steam is routed to the main condenser. Cool water, flowing through the tubes in the condenser, removes excess heat from the steam, which allows the steam to condense. The water is then pumped back to the steam generator for reuse..

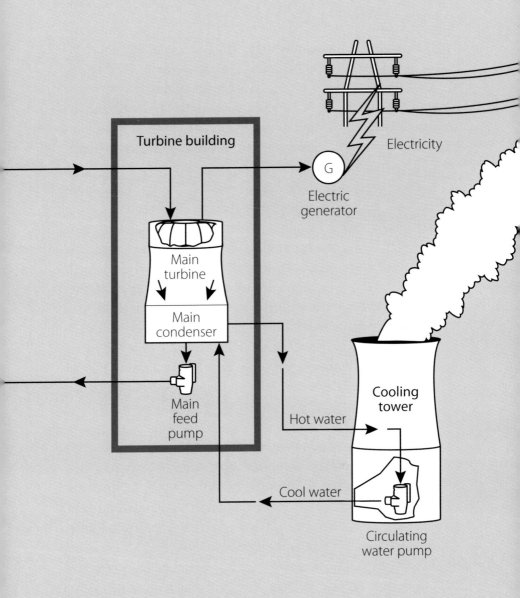

In order for the primary and secondary systems to perform their functions, there are approximately one hundred support systems. In addition, for emergencies, there are dedicated systems to mitigate the consequences of accidents.

Source: Reactor Concepts Manual: Pressurized Water Reactor (PWR) Systems.
www.nrc.gov.

reactors create more fissionable nuclei than were present in the original naturally fissionable fuel. This process makes more efficient use of the original fissionable material during the lifetime of a nuclear reactor core and extends the life of the reserves of the original fissionable nuclear fuel in general. But breeder reactors are expensive to build and operate, and they present dangerous hazards of their own, including the potential for release of highly toxic radioactive waste into the environment.

Supporters of nuclear power see it as a relatively safe alternative to fossil fuels such as coal, oil, and gas. Operating a nuclear plant does not release carbon dioxide, although the construction of such plants and the mining of their fuel do. Nuclear power carries other risks, however, including an uncontrolled meltdown that might release radiation into the environment. Other dangerous scenarios include the use of nuclear material by a terrorist group to manufacture a nuclear weapon; or a leak in a nuclear facility that endangers public health over a wide area, such as that at Chernobyl, Ukraine, in 1986. As a result, the use of nuclear power as an alternative to fossil fuels is still a matter of debate.

Solar, Wind, and Geothermal Power

Coal remained the most important fuel source for power generation through the first half of the twentieth century. But burning coal for power created environmental problems: acid rain, smoggy skies, and toxic emissions that were hazardous to health. When the Clean Air Act and other new laws began controlling the use of coal and other fossil fuels, power companies began to consider alternative sources. The drive for renewable energy intensified as the public grew aware of greenhouse gas emissions and the global warming created by the use of fossil fuels.

In support of the Arabs in the Arab-Israeli War in 1973, members of the Organization of Petroleum Exporting Countries (OPEC) placed an embargo on oil exports to the United States to pressure them to stop supporting the Israelis and to motivate

Australia's Spray-On Solar Cells

One of the major problems with generating solar energy is the initial cost of photovoltaic cells (PVCs), the devices used to convert sunlight directly into electricity. The conventional manufacturing process for photovoltaic cells requires them to be coated with silicon nitrate, which reduces reflection and improves the efficiency of the panels. The demand for solar energy infrastructure remains high and is increasing throughout the world. As a result, manufacturers are seeking out new ways to reduce their costs and become more competitive in a lucrative market. In Australia, engineers are working on a simple and inexpensive solution. Instead of applying the silicon nitrate coating in a vacuum, a new antireflective film is sprayed onto the cells while they travel down a conveyor belt. The equipment and the process are far less costly than the traditional method. According to a February 12, 2009, report titled "Spray-on Solar Panels," posted online by *Alternative Energy News*, the spray-on technique could save a new factory $5 million in the cost of new equipment. The new technique could be available by the end of 2011. If successful, it will reduce worldwide the expense of generating power from the sun. Australian companies will be poised to compete in a rapidly expanding global market for solar-power capacity. "The astonishing fact," reports *Alternative Energy News*, "is that presently the global market for solar cells is growing at a faster rate than markets for mobile phones, digital cameras and laptops!"

them to get Israel to relinquish captured Arab lands. This embargo raised the price of crude oil, as well as fuel for automobiles and electricity-generating plants. The embargo eventually ended, but the energy market became aware that supplies of this vital fuel source were subject to the whim of foreign governments—not all of them friendly to the United States. Interest in renewable energies increased—especially in solar energy.

The development of the photovoltaic cell (PVC) in the 1950s made possible the generation of power from sunlight. Modern PVCs come in the form of flat panel arrays, which use a layer of semiconductor material such as silicon to transform light energy into electrical current. These panels are expensive to manufacture and to install. Once up and running, however, solar energy benefits from an unlimited fuel source, which can be interrupted only by darkness or cloud cover. The production of electrical power by solar energy has steadily increased in the first decade of the twenty-first century. Many industries now derive a share of their electrical power from solar installations on site; when the solar mechanism is producing insufficient energy, factories have the option of turning to the older electrical grid for their power. Many companies have made purchase agreements with their local electrical utilities. These agreements allow a company to sell any unused solar power back to the grid in times of excess supply.

Solar power has become more economical for homeowners as well. The federal government, and local governments as well, are offering tax breaks for homeowners who install solar and other renewable energy systems. In addition, local electric companies offer homeowners the option of selling their extra solar- or wind-generated power, a system that helps the utility to meet peak demand during hot summer days, when air conditioning systems are used to their full capacity.

Wind power is also developing as an important power source in many states. To capture energy from the wind, power companies build turbines that resemble tall windmills. Turbines vary in their size and capacity; the largest turbines in 2010 generated up to 7.5 megawatts of electricity, depending upon the wind speed, which is variable and consequently produces variable power output. Wind-energy companies gather the turbines in large wind farms in places where the wind is strongest and steadiest: a few miles offshore, for example, or on a high elevation.

Wind power requires a large initial investment in equipment as well as land. In Europe, most governments are subsidizing new windpower generation. Denmark has become the leader in this effort, having built enough capacity by 2010 to generate almost 20 percent of its energy needs from wind.

Producers of wind power, however, have several problems to solve. Turbines are noisy and unsightly, and most communities can resist installation of new turbine farms with zoning regulations that prevent their construction. Also, because wind is often intermittent, power generation does not always match demand, and wind energy cannot be stored. This limitation means a backup system is necessary in times of high demand or low fuel (wind) supply.

Foreign governments have invested in wind power as a way to meet the demands of the Kyoto Protocol, which sets binding national limits on the emission of carbon dioxide and other greenhouse gases (the United States has not ratified the treaty). Generating electricity from wind creates no greenhouse gases or pollution of any kind. Wind is a "green" energy that may surpass solar as a leading renewable power source in the future.

The power industry has also turned to geothermal energy, which draws on the heat stored beneath the surface of the earth. Geothermal plants are located above active faults in the earth's crust that allow heat to escape. The geothermal energy resources available to the power industry are vast, and this form of renewable energy has the potential to largely replace the use of fossil fuels.

Geothermal energy can be used to generate electricity or for direct heating of homes, office buildings, and factories. A system of heat exchangers and pipes can send hot water directly to interior radiators, or it can turn the blades of an electrical turbine in a generating plant. For large-scale power generation, however, geothermal energy requires a large investment in equipment for drilling, capturing, and transmitting heat.

Case Study: Germany's Mandatory Approach to Renewable Energy

Although some countries encourage renewable-energy projects with tax incentives and public investment, others mandate by law the development of green energy. In 1998, the Green Party of Germany won an important election, and several of its members joined the country's leaders in an effort to adopt a nationwide program in favor of renewable energy. In support of this effort, the German legislature passed a law mandating that industries obtain their energy from any available renewable sources before turning to fossil-fuel sources. In addition, German households that produce their own power from renewable sources have the option to sell that electricity into the national grid at a fixed price. A law passed in 2001 also required the gradual phaseout of nuclear energy production in Germany by 2020.

Germany and other countries are acting at the direction of the European Council, an organization in Brussels, Belgium, that sets down rules and restrictions for every member of the

German chancellor Angela Merkel (second from right) examines a model of the world's first power station to burn coal without carbon dioxide emissions. AP Images/Sven Kaestner.

European Union. In 2007 the council adopted a green-energy plan for the continent that aims to reduce carbon dioxide emissions by 20 percent by 2020 and have renewable sources provide 20 percent of total power output. With its green laws, Germany has already reached a renewable-source figure of 16 percent and is well on its way to meeting the target.

In Germany, the law requires electric utilities to accept power from and pay a fixed price to those companies and households that are providing energy to the grid from renewable sources. This feed-in tariff is paid for solar, water, wind, geothermal, biomass, and renewable gas sources such as landfills and sewage-treatment plants. It supports and encourages green-energy projects, as investors know in advance what they will earn by providing power. The feed-in tariff decreases gradually year by year, providing an incentive to the renewable-power industry to improve efficiency and lower its costs.

Not everyone in Germany supported the law when it was originally passed. Electric companies saw it as a burden that would raise their costs unnecessarily. Some consumers and businesses also protested, realizing that the costs of the feed-in tariff would be passed on to them. As the feed-in tariff has declined and renewable energy has become gradually less costly, however, the law has gained widespread acceptance. Germany accounts for about half of new solar-energy installation globally; its photovoltaic arrays provided 3.8 gigawatts of power in 2009, a rate that has made Germany the largest market for photovoltaic cells in the world.[1]

Despite its government mandates and supports, Germany will be encountering a shortfall in energy generation with the mandated phaseout of nuclear plants. To replace nuclear plants that are taken out of commission, German energy companies are planning more than two dozen new coal-fired power plants, which will draw on inexpensive coal mined in neighboring Poland and other countries. One such plant, to be built near the German capital of Berlin, will generate 800 megawatts of power and burn up to two million tons of coal a year. The plant will also generate five million tons of carbon dioxide emissions yearly, making it one of the country's largest sources of greenhouse gas emissions.[2]

Although these new plants will be more efficient than older coal-fired plants, they may prevent Germany from reaching its goal of a 20 percent reduction in greenhouse gas emissions by 2020. To help meet emissions targets, the new coal plants will rely on carbon capture and storage technologies. But these technologies remain in the developing stage, and by 2010 they were still not significantly reducing the amount of carbon dioxide released into the atmosphere by existing plants.

Yet the new coal plants are supported by political leaders such as Germany's chancellor Angela Merkel, who is a leading advocate of green-energy projects in Europe. Merkel is seeking to encourage renewable energy but also wants the investment and jobs that the new coal plants will bring. The sentiment against nuclear power in Germany is high, and Merkel's government steadfastly supports the phaseout of nuclear plants. As a result, without new coal plants Germany will have a serious energy shortfall and become more dependent on natural gas imported from Russia, a resource that is threatened by any political or economic friction between the two countries. In this way, coal provides a measure of energy security for a nation still dependent on foreign supplies for its power.

Clean Coal and the Schwarzenegger Clause

In 2006 the state legislature of California passed SB 1368, a law that set down new greenhouse gas emission standards for the state's public utility companies. Although the state used a baseline that matched the average emission level of natural-gas power generation, the target of the new law was coal. Any new coal-fired plant in California would have to limit its emissions to 1,100 pounds (500 kg) of carbon dioxide for every megawatt-hour of electricity that it generated, a benchmark that came to be named the Schwarzenegger Clause after California's governor Arnold Schwarzenegger, who threw his support behind the law.

A sponsor of the law, Senator Tom Perata, commented that "California enacted SB 1368 to send a strong signal to the western energy markets. Our energy must be clean—we won't buy power from coal plants spewing greenhouse gases by the ton."[3] Following the passage of this law, the states of Washington and Maine also adopted the standard; the European Union has proposed that any new coal plant built in its member countries meet this standard after 2015.

The Schwarzenegger Clause means that new coal plants will have to take up new clean-coal technologies in order to operate within the law. Yet carbon capture and sequestration and other methods of reducing emissions from coal are still in an experimental stage. If the law remains in effect, new coal plants will remain on the drawing board, and power companies in California and other Schwarzenegger Clause states will have to turn to natural-gas-fueled plants, to renewable energy sources, and to nuclear power.

Notes

1. "Germany's Amended Feed-In Tariffs Favour Home Solar Power," Energy Matters, July 12, 2010. www.energymatters.com.au.
2. Roland Nelles, "Germany Plans Boom in Coal-Fired Power Plants—Despite High Emissions," Spiegel Online, March 21, 2007. www.spiegel.de.
3. SourceWatch.org, "Schwarzenegger Clause." www.sourcewatch.org.

CHAPTER 6

Solving the Ozone Hole: The Montreal Protocol

Synthetic chemicals have played an important role in the industrialization of the modern world, and in the invention of new consumer products. Freon, for example, is a synthetic gas used in air conditioners and refrigerators. Plastic, nylon, and other synthetics are now present in thousands of items that make life easier and more convenient. But chemicals used in industry and manufacturing can also be dangerous to health—not just that of the workers who handle them but also that of consumers who buy the finished products.

Although some nations strictly regulate the use of toxic chemicals, others do not. In the United States, for example, the use of pesticides on food crops is controlled by federal law, which bans some pesticides altogether. The goal of these laws is to prevent toxins from rising to harmful levels in the environment and in the public food supply.

Other substances may be harmful to the public as well but also have a less noticeable effect on the health of individuals who are exposed to them. Individual nations may regulate their use, but if other nations continue to release these toxins into the environment, their impact may be worldwide. In theory, all nations have an interest in controlling the use of such chemicals.

The depletion of the ozone layer by industrial chlorofluorocarbons (CFCs) brought about the first international regulation of chemical compounds. Manufacturing industries fought against the control and banning of CFCs for several years, but eventually the scientific case for regulation overcame the opposition, and an international agreement known as the Montreal Protocol was reached through the United Nations. In addition to being the first international treaty designed to protect the environment, it also remains the most successful.

Freon and Other Useful Compounds

During the middle of the twentieth century, CFCs were considered harmless and useful compounds by the small segment of the population who knew anything about them: chemists and industrial engineers. They were used as cooling agents in air conditioners and refrigerators, and to power aerosol cans. There was little thought given to their effect on public health or the natural environment.

In the 1970s, however, scientists at the University of California began researching the effect of CFCs when released into the atmosphere. They found that CFCs continue to rise until they reach the stratosphere, an atmospheric layer that begins about 6 miles (10 km) above the surface of the earth. CFCs remain intact for several decades until the sun's ultraviolet light eventually breaks them down into several elements. One of these elements, chlorine, in turn breaks down a vital oxygen-based molecule known as ozone.

The presence of ozone in the upper atmosphere is essential to life on earth. Ozone absorbs ultraviolet-B (UVB) rays emitted by the sun that are harmful to humans and other forms of life. If the ozone layer were to be completely depleted, sunlight would cause much greater damage, in the form of skin cancer and other illnesses, to humans. It would also harm food crops and the small plankton that provide a vital source of food in the ocean.

The Growing Ozone Hole and Ultraviolet-B Radiation

The California research brought about a theory, known as the Rowland-Molina hypothesis after the two scientists who devel-

Information and Disinformation

To combat environmentalists and their claims about global warming, companies such as ExxonMobil have funded research and information groups that advertise themselves as independent, nonpartisan, and scientific. The basic strategy of such groups is to publish opinion skeptical of global warming and to promote the ideas of scientists that may clash with the generally accepted consensus, which is that the earth has been warming throughout the industrial age, and human-produced greenhouse gases, including carbon dioxide, are an important cause.

These organizations work to lobby Congress, state legislatures, and the public against new laws and initiatives designed to combat global warming. They fund research that is meant to support their point of view, and they frequently change names in order to disguise their true nature and to appear impartial. In many cases, the groups are guided by legal and public-relations firms that have professional expertise in shaping public perceptions of political candidates and current issues.

In 2008, a group known as Americans for Balanced Energy Choices was transformed into American Coalition for Clean Coal Electricity. In the next year, the group became America's Power Army. The group attempts to garner public acceptance of traditional fossil-fuel use for

oped it, that CFCs deplete stratospheric ozone. When the theory was first published, CFC manufacturers came out strongly against the idea of ozone depletion and any regulation of their industries. The Rowland-Molina hypothesis was also strongly disputed by DuPont, the chemical company that invented CFCs and that stood to lose a significant amount of money if the government ever regulated or banned these compounds. Presenting itself as a responsible corporate citizen, DuPont ran a full-page advertisement in the *New York Times* to proclaim that "Should reputable evidence show that some fluorocarbons cause a health hazard through depletion of the ozone layer, we are prepared to stop production of the offending compounds."[1] The company held its ground against the Rowland-Molina hypothesis through power generation and raise skepticism on the issue of global warming and the development of renewable energy sources. On the Americaspower. org website, the group gives its basic mission statement: "We recognize that coal, America's most abundant energy resource, plays a critical role in meeting our country's growing need for affordable and reliable electricity. Our goal is to advance the development and deployment of advanced clean coal technologies that will produce electricity with near-zero emissions."

In 2006, the Nuclear Energy Institute, which lobbies for the construction of new nuclear reactors, established another front group, the Clean and Safe Energy coalition, with the advice and guidance of Hill & Knowlton, a large public-relations firm. The group emphasizes the advantages of nuclear power on its website, although the word "nuclear" is absent from its title.

Front groups use a variety of tactics to conceal their goals. They might not reveal the source of their funding or their ties to industries or companies that may benefit from their work. They may rename themselves and redesign their websites and publications in order to adopt a more convincing appearance of neutrality. They may also use scientists and other experts as paid "advisors" to publicly support their point of view. The front group has become one of the most useful weapons in the challenge to global-warming science, as well as to the international treaties and government mandates designed to address the issue.

the 1970s, stating that no ozone depletion had been detected and that the entire theory remained uncertain, as it was based on projections and not solid proof.

The scientific evidence for a link between CFCs and ozone depletion grew stronger, however. In the Antarctic polar region, the concentration of UVB reaching the ground was doubling during the seasonal period when the ozone layer was thinnest. Studies found UVB levels rising in Canada, as well, and medical researchers confirmed links between UVB and skin cancer as well as cataracts.

The US government then funded an ozone depletion study by the National Academy of Sciences. Released in 1976, the report supported the Rowland-Molina hypothesis and called for further study of ozone concentration over the polar regions, where depletion seemed to be occurring at the fastest rate. In 1985, British scientists working in Antarctica found what was termed a hole in the ozone layer over that region. The hole's dimensions showed the rate of ozone depletion to be much faster than anticipated.

In the meantime, with the discovery of increasing ozone depletion, DuPont executives realized they needed a more effective response than simply denying the hypothesis. Through the 1980s, the company maintained that it was concerned for public health and safety, and that it stood ready to take action on CFCs if any solid scientific proof should appear that supported the case against them. To organize its response and public communications, the company formed the Alliance for Responsible CFC Policy, an organization that lobbied Congress against any new regulation. DuPont also suspended research into alternatives to CFCs, confident that it had an ally in the business-friendly Republican Party that controlled, at the time, both the White House and the United States Congress.

The Montreal Protocol: Government and Industry Partnership

Traditionally in the United States, private business has remained separate from and often hostile to public government. Since the late 1800s, the federal government has been regulating corporations in the stated interest of protecting the public from monopolies and trusts (which allow business leaders to control the supply and the price of essential goods). Following this tradition, DuPont and other corporations strongly resisted any control of their business by any form of government law or regulation. Such regulation, it was believed, would reduce their income by making it more expensive to manufacture and market their products.

The scientific case for ozone depletion by CFCs grew stronger through the 1980s, however. The concentration of CFCs above the earth was increasing. The ozone layer was thinning out every year, recovering, and then thinning again. As CFC concentrations increased, it seemed possible to many that in certain regions of the earth, protective ozone could vanish completely, leaving surface regions unprotected against the sun's ultraviolet light. These dangers would be present around the world, and not just in those countries that manufactured CFCs. It became apparent that the only effective mechanism to control CFCs would be some form of international agreement.

In the same year as the discovery of the ozone hole over Antarctica, a group of twenty nations met in Vienna, Austria, to address the problem. Although banning or regulating CFC production could not be accomplished immediately, the representatives did agree on a framework that would allow the creation of a multilateral agreement. After the representatives signed the Vienna Convention for Protection of the Ozone Layer in 1985, the United Nations took up the issue. In 1987, in Montreal, twenty-four member countries of the United Nations signed the Montreal Protocol on Substances That Deplete the Ozone Layer.

The treaty set specific goals for the reduction of ozone-depleting chemicals, including a reduction of their use by 50 percent by the year 1999. Further amendments to the agreement provided for a gradual phaseout of CFC production over a period of years. This amendment allowed DuPont and other companies time to develop environmentally friendly alternatives. The protocol was signed by President Ronald Reagan and then sent to the United States Senate, which must ratify all international treaties.

Protest, Then Cooperation by Industry

Skeptics of the ozone depletion theory remained active, vocal, and strongly opposed to the treaty the president had signed. One of their leaders in the media was radio commentator Rush Limbaugh, who proclaimed that "Mount Pinatubo in the Philippines spewed forth more than a thousand times the amount of ozone-depleting chemicals in one eruption than all the fluorocarbons manufactured by wicked, diabolical and insensitive corporations in history. . . . Mankind can't possibly equal the output of even one eruption from Pinatubo, much less 4 billion years' worth of them, so how can we destroy ozone?"[2]

DuPont continued its fight to protect CFC production. The Senate called the company's leaders and experts to Washington to hear their testimony. Before the legislators, DuPont's representatives claimed that no crisis was at hand, and that there was still no scientific justification for regulation or banning of CFCs. In 1988, Chairman Richard Heckert of DuPont wrote to a group of senators, stating that "At the moment, scientific evidence does not point to the need for dramatic CFC emission reductions. There is no available measure of the contribution of CFCs to any observed ozone change."[3]

Nevertheless, the Senate ratified the Montreal Protocol, which became the law of the land in the United States and eventually in the 196 other countries that signed the agreement. Under the London Amendments passed in 1990, DuPont and other companies would also have to cease manufacturing halons and carbon tetrachloride—other damaging chemical compounds—by 2000. Under further amendments made in Copenhagen in 1992, CFCs would have to be completely phased out of production by 1995.

Alternatives to Chlorofluorocarbons

Eventually 196 nations—every member of the United Nations—would sign the treaty. In the United States, the Clean Air Act of 1990 implemented the Montreal Protocol and the US Environmental Protection Agency (EPA) became the federal watchdog overseeing use of ozone-depleting chemicals. The manufacture of CFCs represented a multibillion-dollar industry, but the law allowed DuPont a few years to create and market an alternative. The company took full advantage of the opportunity by turning to a new family of chemical compounds, known as HCFCs (hydrochlorofluorocarbons) and HFCs (hydrofluorocarbons). These compounds can be used in refrigeration systems, in air conditioners, and as propellant in fire extinguishing systems. Although they have less potential to cause harm to the ozone layer, they are potent heat-trapping gases when released into the atmosphere. According to one report, one type of HCFC lasts 52 years in the atmosphere and has 4,470 times the global warming impact of carbon dioxide.[4]

The United Nations had solved an environmental problem but created an economic one. Many countries did not have large chemical industries of their own and would be hard-pressed to develop or manufacture CFC alternatives. The demand for refrigeration and air conditioning systems in these nations was growing along with their economies. For this reason, the UN also created an international fund, which would help developing countries pay for the phaseout of CFCs. Money in the fund is donated by various nations, including the United States, and helps developing countries adopt CFC alternatives and pay for the invention and development of these new chemicals.

A recycling plant removes toxic chemicals from an old refrigerator, including chlorofluorocarbons, deemed responsible for damaging the ozone layer. AP Images/PRNewsFoto/JACO Environmental, Inc.

The Current State of the Ozone Layer

In a report published in 2006, a team of scientists at the National Oceanic and Atmospheric Administration found that climate change as well as ozone-depleting gases were still playing a role in the depletion of ozone: "The Antarctic ozone hole now is not

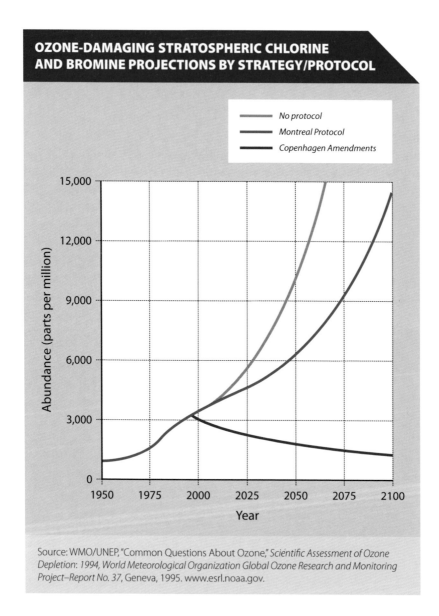

OZONE-DAMAGING STRATOSPHERIC CHLORINE AND BROMINE PROJECTIONS BY STRATEGY/PROTOCOL

Legend:
- No protocol
- Montreal Protocol
- Copenhagen Amendments

Y-axis: Abundance (parts per million) — 0, 3,000, 6,000, 9,000, 12,000, 15,000

X-axis: Year — 1950, 1975, 2000, 2025, 2050, 2075, 2100

Source: WMO/UNEP, "Common Questions About Ozone," *Scientific Assessment of Ozone Depletion: 1994, World Meteorological Organization Global Ozone Research and Monitoring Project–Report No. 37*, Geneva, 1995. www.esrl.noaa.gov.

as strongly influenced by moderate decreases in ozone-depleting gases, and the unusually small ozone holes in some recent years (e.g., 2002 and 2004) are attributable to dynamical changes in the Antarctic vortex. The anomalous Antarctic ozone hole of 2002 was manifested in a smaller ozone-hole area and much smaller ozone depletion than in the previous decade. This anomaly was due to an unusual major stratospheric sudden warming. . . ."[5]

In more recent studies, scientists have found that the concentration of CFCs in the upper atmosphere has decreased, and that the ozone layer has begun to recover. By some measures, protective ozone in the stratosphere is expected to return to its natural level by the middle of the twenty-first century. According to a conclusion of the *Scientific Assessment of Ozone Depletion 2010*, prepared by the World Meteorological Association:

> The Montreal Protocol is working. It has protected the stratospheric ozone layer from much higher levels of depletion by phasing out production and consumption of ozone-depleting substances (ODSs). Simulations show that unchecked growth in the emissions of ODSs would have led to ozone depletion globally in the coming decades much larger than has been observed. Solar UV-B radiation at the surface would also have increased substantially.[6]

The Problem of Replicating a Montreal Protocol for Carbon Emissions

Before the Montreal Protocol, environmentalism and business seemed to be, inevitably, at odds. Environmental protection was a zero-sum game, in which a win for one side meant a loss for the other. Business leaders felt their companies were threatened by any limits on what they could produce or sell. Nations that wanted to protect their water, air, and soil from toxic hazards had to sacrifice economic growth. Others that sought rapid development would have to allow toxic chemicals to foul their environments.

To some degree, the Montreal Protocol showed otherwise. It was the first international agreement designed to counter an immediate environmental threat. By helping developing nations replace useful but ozone-depleting chemicals, the treaty also addressed the balance of economic and environmental interests. The success of the treaty has shown that international cooperation on environmental issues is possible if the agreement solves economic concerns of developing nations.

For most nations, the economic well-being of the population takes priority over vague threats to the natural environment in their own countries or abroad.

International agreements that do not balance economic and environmental concerns, however, will fail. For most nations, the economic well-being of the population takes priority over vague threats to the natural environment in their own countries or abroad. If development is held back and large segments of the population remain in poverty, governments will lose public support, so it becomes inexpedient both economically and politically to support environmental objectives and proposals.

As a success, however, the Montreal Protocol provided a model for future agreements, particularly the Kyoto Protocol, which binds the countries that sign it to limits on greenhouse gas emissions. But the Montreal Protocol addressed a problem that was fairly easy to solve: the use of a single family of chemical compounds that could be replaced. The Kyoto agreement took on a much larger and more complex problem, and it has met with very limited success.

Notes

1. *New York Times*, June 30, 1975.
2. Quoted in *Fairness and Accuracy in Reporting*, "The Way Things Aren't," July/August, 1994. www.fair.org.

3. "DuPont: A Case Study in the 3D Corporate Strategy," Greenpeace Position Paper, September 1997. http://archive.greenpeace.org.
4. David A. Fahrenthold, "Chemicals That Eased One Woe Caused Another," *Washington Post*, July 20, 2009. www.washingtonpost.com.
5. "Scientific Assessment of Ozone Depletion: 2006; Executive Summary," National Oceanic and Atmospheric Administration, March 2007. www.esrl.noaa.gov.
6. "Scientific Assessment of Ozone Depletion: 2010," United Nations Environment Program, September 2010. http://us-cdn.creamermedia.co.za.

CHAPTER 7

The Kyoto Protocol: International Limits on Carbon Emissions

In the twenty-first century a company founded in one nation may cross a border to invest in another, and to set up plants to take advantage of cheaper labor, lower taxes, or lighter regulation of its industry. Large energy companies, for example, sign leases with governments to search for natural gas and crude oil; developing countries welcome foreign automakers to build factories and hire new employees. Jobs and growth are a priority for governments seeking to win and keep the support of their populations.

Developed countries, such as the United States, Japan, and the nations of western Europe, enjoy higher standards of living, but they still compete with poorer nations for investment. Although the developed nations have stricter environmental laws, they approach international treaties, such as the Montreal Protocol, with great caution. A phaseout of chlorofluorocarbon (CFC) production, for example, can mean higher production costs when alternative chemicals have to be developed. This drives new investment to developing countries with a generally lower cost of business.

The control of greenhouse gas emissions is a much tougher environmental problem. The release of carbon dioxide is an integral part of modern life. Manufacturing and transportation all consume energy and release carbon dioxide. Consumers buy and use products that require carbon-intensive industries; the more developed a country is, the faster its people emit greenhouse gases.

In nearly all nations of the world, economic interests take priority over environmental worries. Developed countries such as the United States see their industries and living standards threatened by treaties that limit carbon emissions. For this reason, developed countries seek to have developing countries included in any agreements made on global warming and greenhouse gas emissions. Developing countries, on the other hand, believe that limiting their emissions and thus their economies harms their people and puts them at an unfair economic disadvantage, and so they seek to be exempted from the limitations imposed by these treaties.

A hot climate will not discriminate: All nations will be affected, one way or another.

For their part, environmentalists see preservation of a livable global climate as the priority. Leakage is their term for the movement of carbon-intensive industries from developed to developing nations. From their point of view, if leakage negates the effect of global treaties, and global warming gets out of hand, then economies around the world will suffer. A hot climate will not discriminate: All nations will be affected, one way or another, by rising temperatures and sea levels, more frequent extreme weather events, higher rates of disease, declining food production, and the many other possible consequences of global warming.

Individuals supporting international agreements point out that environmental regulation need not harm business. In the United States, a trading mechanism in emission allowances helped power companies survive the Clean Air Act of 1990. In addition, new regulation of emissions creates a need for new products with a smaller greenhouse gas impact, or carbon footprint. The market for green products and technology will reward innovation, just as the demand for automobiles created some of the most profitable industrial companies in history.

The Kyoto Protocol

In 1992 the United Nations held a Conference on Environment and Development in Rio de Janeiro, Brazil. The nations meeting at the Rio Earth Summit discussed the conflict between economic development and environmental protection. At the end of the conference, the participants signed the Rio Declaration and the Framework Convention on Climate Change (FCCC). The 154 countries that signed the FCCC legally bound themselves to follow its directives.

The FCCC had no specific targets for reductions in carbon dioxide and other greenhouse gases. Nor did it set up time frames for compliance or penalties for violators. Instead, it set a vague but ambitious goal: stabilization of greenhouse gas emissions, while allowing economic development. The agreement also obligated developed countries to set an example in taking positive steps to reduce their emissions. To implement the agreement, the signing countries agreed to meet in regular conferences in order to devise methods of reaching their goal. Conferences were duly held in Berlin and Geneva; a third conference took place in Kyoto, Japan, in 1997.

The meeting of the nations at Kyoto turned out to be a major turning point in the global warming debate. The result of the meeting was the Kyoto Protocol, the first international agreement that bound the countries that ratified it to reduce their greenhouse gas emissions. Kyoto set a goal of reducing greenhouse gas emissions during a commitment period that would begin after the treaty was ratified by enough countries to put it into effect. Every country that signed and ratified the agreement would be pledging to achieve the emissions targets that Kyoto set down. To achieve these targets, Kyoto allowed the use of a carbon-emission trading program (also known as cap and trade) and several other mechanisms. The agreement was scheduled to go into effect once countries representing 55 percent of global greenhouse gas emissions ratified it.

In the United States many business leaders opposed Kyoto, although President Bill Clinton supported it. With the election of George W. Bush to the presidency in 2000, opponents gained an ally in the White House. At the time of the election, the Kyoto Protocol had still not gone into effect, nor had the United States Senate ratified the agreement. Bush announced his opposition to Kyoto, and with his Republican Party also in control of the Senate, there was little chance of the United States ratifying the agreement. With its large share of global emissions, the United States was seen by nations that had signed Kyoto as a vital partner in the process. American politicians continued to wrangle over the treaty, however, and President Bush maintained his opposition through the election campaign of 2004, which ended with the president's election for a second term.

In the next year, the ratification of Kyoto by Russia finally set the agreement into effect. The Kyoto nations had to monitor their emissions and submit detailed reports to the United Nations, which set up a bureau to keep track of compliance. In addition, all carbon trading was recorded and monitored by a United Nations Climate Change Secretariat, located in Bonn, Germany.

Annex I Countries and Emission Targets

The Kyoto conference drew up a list of Annex I countries that included twenty-eight members of the developed world that had played the major role in creating the global-warming problem, and now would play a leading role in helping to solve it. In 1998 Kyoto created an Annex II by adding several countries that were designated as economies in transition or, in other words, on the way to developed status. These new nations included Slovenia, Romania, Ukraine, and other nations in eastern Europe, which had recently thrown off Communist economic systems and were adopting a capitalist system similar to that of western Europe and the United States.

Kyoto's ultimate goal was a global reduction in the emission of greenhouse gases in the amount of 5.2 percent below 1990 levels. The amount of the cuts proposed by the agreement varied: 8 percent in European Union countries, 7 percent in the United States, and 6 percent in Japan.

Developing countries, including Bangladesh, India, and China, were excused from Kyoto's cuts and restrictions. Although they might emit as much greenhouse gas, per capita, as the Annex I and II nations, Kyoto recognized that limiting their emissions and thus their economies in any way would place an unfair burden on them, as these developing countries had played a much smaller role in the global-warming problem.

By 2005, the year the Kyoto Protocol went into effect, China was producing 17 percent of the globe's greenhouse gas emissions, the largest share of any country. By 2009 the Chinese were approaching the per-capita emission levels of western Europe. Whereas each person in the United States was responsible for an average of 17.2 tons of greenhouse gas emissions and each European Union resident was responsible for an average of 7.9 tons, each person in China had, on average, a 6.1-ton carbon footprint, a number that was steadily rising through the first decade of the twenty-first century.[1] The Chinese government, in the meantime, strongly resisted any attempt to place limits on its carbon emissions. Officials believed that the voluntary limits placed on it by Kyoto would allow the country to develop green technology and industry in the most efficient way.

Kyoto Mechanisms

The Kyoto Protocol established a complicated quota mechanism intended to reduce greenhouse gas emissions. Each country was given assigned amount units or AAUs (a single AAU represents the emission of a ton of carbon dioxide). With its AAU quota, each country then grants allowances or carbon credits to individual companies within its borders. The companies can buy the credits if they are exceeding their emission quota or sell the credits if they are emitting less and have allowances to spare. This system gives companies an incentive to invest in clean energy projects that would help them meet their targets and earn income from the sale of their credits.

The carbon credits, similar to allowances created by the 1990 Clean Air Act in the United States, can be traded on an international market. In addition, removal units (RMUs) can be created on the basis of land use and the management of forests. This provision of the Kyoto treaty takes into account the fact that forested areas are carbon sinks that absorb, rather than emit, carbon dioxide. Each RMU equals a metric ton of carbon dioxide, removed from the atmosphere by the natural process of plant photosynthesis. The removal units can also be sold, providing a source of income in countries that reforest large areas under development or cultivation.[2]

Finally, individual companies can earn offsets by developing clean energy projects that help reduce greenhouse gas emissions. A power plant drawing on solar or wind energy, for example, can be certified as a clean development mechanism (CDM) project. Another example of a CDM project is the replacement of outdated factory machinery with equipment that reduces or eliminates the emission of carbon dioxide. These projects can be carried out in the home country or abroad; they earn a certified emission reduction for each ton of carbon dioxide they remove from the atmosphere. The country where the project takes place must have signed and ratified the Kyoto Protocol.

The clean development mechanism is limited to projects taking place in non–Annex I countries. The goal is to promote sustainable development: the building of self-sufficient industries that do not contribute to global warming. CDM allows companies in developed countries to carry out green projects in developing nations; the carbon credits earned can then be applied to factories and properties in the home country. The CDM would help poorer countries attract new investment, stimulate local economies, provide jobs, and earn a profit for the sponsor.

The joint implementation (JI) program, by contrast, takes place only in Annex I countries. JI is a way for developed countries to take advantage of the carbon credit system to build or adopt green-energy projects. The carbon credits earned can be applied to a country's carbon budget and sold or traded on the international market.

Developed Versus Developing Countries

As Kyoto went into effect, and countries began regulating their industries to comply, the split between developed and developing nations remained. Many political leaders in the Annex I countries believed the agreement unfairly punished them by placing limits on their industries. Developing nations, such as China, Indonesia, Mexico, and India, asserted that the advanced, wealthier countries bore more responsibility for the global warming problem, and therefore should sacrifice more to help solve it.

At a climate conference held in Bangkok in 2008, a sharp debate over the responsibilities of developed and developing countries ensued. Representatives from the United States asked for a new approach. Instead of directing emissions cuts on a fixed international timetable, this approach would allow each nation to pledge emissions reductions according to its own timetable. According to supporters, this was a more realistic approach that would help both developed and developing countries assist in the effort while taking into account their very different abilities and stages of economic development.

Diplomats from China and other developing nations opposed this idea, however, and accused the developed nations, including the United States, of trying to sabotage the Kyoto Protocol and prevent any future agreement on greenhouse gas emissions. In their view, the flexible approach was merely a way of allowing developed nations to escape any effective limits on their emissions and to place more responsibility on the developing nations. China, India, and other nations pointed to the failure of the US Senate to ratify the Kyoto agreement, and claimed that the United States had abandoned its moral duty to lead, or at least support, the international effort on climate change. These developing countries want to keep the basic Kyoto framework: setting down legally binding emissions cuts, based on a generally accepted scientific consensus, on a specific timetable.

The opposing view holds that by setting their own targets, developed countries can still use carbon trading, emission allowances, and the clean development mechanism to reach reasonable goals. Limitations and penalties on established industries, which would directly affect economic output, would not be necessary. This view also holds that developing and developed countries must share equal responsibility in the effort, however. According

A Bangladeshi worker carries clay at a factory in the outskirts of Dhaka. The Kyoto Protocol excused developing countries such as Bangladesh from restrictions on emissions to avoid hampering their economic activity. AP Images/Pavel Rahman.

to Martin Khor, an observer at the Bangkok conference, "The United States wants only to have a national target without binding it to a global treaty. . . . They are stressing that developing countries have 'common' responsibilities, a code for pulling in the developing countries into emission-reduction obligations, while down-playing the 'differentiated' responsibilities that recognise that the developing countries have had little role in the historic emissions and need space for economic development."[3]

The Political Fight over Ratification

Although Kyoto was eventually ratified by most countries that signed the agreement, the treaty ran into strong opposition in the United States, where international agreements have been rejected, delayed, or amended by the legislative branch of government ever since the country's founding in the eighteenth century. In a 2001 *New York Times* opinion piece, commentator Walter Russell Meade notes:

> The politics of American foreign policy has been divided on regional and cultural lines at least since the South and the West forced the War of 1812 on a sullen New England. The Senate and the White House have been squabbling over foreign policy since George Washington, stung by sharp Senate questioning over a proposed Indian treaty, decided he would never return to seek its advice and consent in person.[4]

In the traditional view of business, the economic interests of the country are more important than cooperation with foreign nations, and political leaders strongly resist any agreement that limits or regulates American industry.

Kyoto was one such agreement. Without the participation of the United States, however, which emits a large share of the earth's greenhouse gases, many believed the treaty would be ineffective. Lord May, a British scientist and government adviser, calculated that the rise in emissions from the United States alone would be greater than the cuts achieved by Kyoto, even if every country that participated reached its goal. "Even if emissions from the United States stay at the same level until 2012," May commented, "which is an unrealistically conservative assumption, while the other targets are met, the overall results for the original parties to the Protocol will be a rise in emissions of 1.6 per cent instead of the desired reduction of 5.2 per cent."[5]

Nevertheless, many politicians in the United States were not persuaded by this argument. The treaty would, in effect, force industries to accept higher energy costs, which would be passed along to people who purchased their goods, contributing to inflation and a lower standard of living. As a result, the treaty would have an unacceptable, negative impact on the United States and on any country that signed it, an opinion voiced even by political leaders in nations that signed the agreement. Vice President Al Gore, a prominent leader of environmental efforts in the United States, voiced his concern that the agreement did not put in place a mechanism for controlling emissions from developing nations; for that reason, the administration of President Bill Clinton never sent the agreement to the US Senate for ratification. Andrei Illarionov, an economic minister from Russia (where ratification put the agreement into effect), proclaimed that "The Kyoto Protocol is a death pact, however strange it may sound, because its main aim is to strangle economic growth and economic activity in countries that accept the protocol's requirements."[6]

Opponents also believed industries in Kyoto-signing nations would suffer a competitive economic disadvantage. Industries in countries that did not ratify the treaty would benefit from lower costs, and would be able to undercut prices charged by their competitors in Kyoto-signing nations. Investment and jobs would migrate across borders, thus leaving signatory countries with stagnant economies and rising unemployment.

Kyoto, opponents also pointed out, raised conflicts over international trade agreements, such as the North American Free Trade Agreement (NAFTA). Countries that sign such agreements pledge not to raise taxes, quotas, or limits on products that are imported

Kyoto Pros and Cons

The controversy over the Kyoto Protocol has produced reams of commentary from supporters and opponents, and very little from individuals who feel uncertain of their positions and the treaty's ultimate outcome. A small sample of the many debating points follows.

Should Developing Countries Contribute?

"There are some who will try to pervert this precedent and use xenophobia or nativist arguments to say that every country should be held to the same standard. But should countries with one-fifth our gross domestic product—countries that contributed almost nothing in the past to the creation of this crisis—really carry the same load as the United States? Are we so scared of this challenge that we cannot lead?"

—*Al Gore, "Moving Beyond Kyoto,"* New York
Times, *July 1, 2007*

"Any agreement that allows the developing countries to continue emitting greenhouse gases would in effect negate the efforts of those countries that are trying to reduce them."

—*Charli Coon, "Why President Bush Is Right to
Abandon the Kyoto Protocol," Heritage Foundation,
May 11, 2001*

Will Kyoto Have a Negative Economic Impact?

"Kyoto is among the least cost-effective ways to address climate change. It thus violates the UN framework convention on climate change (to which both Britain and the US are parties), which obliges governments to pursue 'cost-effective' measures to address climate change. Our poll

from other countries, yet the costs of complying with Kyoto might make such barriers necessary. These limits could also be applied to the emission allowances that companies were supposed to trade with each other in order to comply with Kyoto's limits.

found that 70% of Britons believe that the UK should pursue these more cost-effective measures, rather than Kyoto."

—Kendra Okonski (letter), "The Case Against Kyoto,"
The Guardian (UK), February 21, 2004

"It's a false choice to say we need to favor the economy over the environment. Especially given the progress we've made in developing the technological know-how to profit from a shift to cleaner energy production. The previous Administration was successful in working with business and environmental groups toward an agreement that protects both American interests and the world environment. The Bush Administration would have been wise to build upon this success."

—Senator Joe Biden, letter to President George W.
Bush, 2001

Should the United States Join the Kyoto Agreement?

"Signing the Protocol, while an important step forward, imposes no obligations on the United States. The Protocol becomes binding only with the advice and consent of the U.S. Senate. As we have said before, we will not submit the Protocol for ratification without the meaningful participation of key developing countries in efforts to address climate change."

—Vice President Al Gore, "Statement on Signing the
Kyoto Protocol," November 12, 1998

"I walked away from Kyoto because it would have damaged the American economy, it would have destroyed the American economy. . . . It was a lousy deal for the American economy."

—President George W. Bush, quoted in Forbes.com,
July 4, 2005

The various mechanisms of Kyoto have also come under fire. In order for the clean development mechanism to work, for example, there must be a strict and accurate accounting of emissions both before and after a company puts a green project in place. If emissions cannot be accurately recorded, the company may still earn credits simply by planning the project and having it certified by the appropriate government agency. This system can give rise to dishonest accounting—motivated by bribery and graft—on the part of company officials and government ministers, which thwarts Kyoto's goal of efficiently reducing greenhouse gas emissions.

Status of Emission Reductions and the 2012 Deadline

The Kyoto Protocol set down the years 2008–2012 as the period in which its emissions targets would be achieved. As of 2010, however, the promise of the treaty has been largely unfulfilled. The original Annex I countries are, collectively, emitting more greenhouse gases in 2010 than they were in 1990, instead of less. Canada, Australia, and New Zealand have seen significant increases, mainly because these economies have been growing faster than other nations affected by a global recession. The two nations belonging to Annex I that have not ratified Kyoto, Turkey and the United States, have raised their emissions at a faster rate, on average, than nations that signed the agreement.

The new Annex I group (which includes formerly Communist nations in central Europe) is reaching its targets, however, mainly from the effect of an economic transition. These countries, including Russia, have achieved deep emissions cuts not through green projects or carbon-credit trading, but mainly through the phasing out of obsolete heavy industries that operated under their former Communist governments. Unable to earn money in a competitive global economy, these inefficient plants and factories have shut down.

Another Kyoto success has been in the land of the treaty's origin, Japan, which is on track to achieve a reduction of about 5 percent. Japan has implemented several CDM projects in Asia and is also a leading innovator in carbon-storage technology.

The Copenhagen Conference

The Kyoto Protocol will expire in 2012. Supporters of the treaty point out that many European nations are on track to meet their emission targets. Opponents show that Kyoto is still not supported by the United States, and that developing countries still

POLL: WHO SHOULD CUT EMISSIONS?

Respondents from six global economies were asked which group of countries should reduce emissions first.

	Brazil	China	Japan	Russia	United States	Germany
Countries like the United States, Germany, and Japan	22%	26%	21%	13%	10%	16%
Countries like China, India, and Brazil	7%	13%	10%	6%	13%	15%
Or, both at the same time	55%	44%	66%	60%	75%	68%
None/ Don't Know/ Refused	16%	17%	3%	22%	2%	1%

Source: *Gallup*, 2009.

do not see binding limits as a just way of solving the problem. At the Copenhagen climate conference held in 2010, China steadfastly refused to change its opposition to mandated emission limits, a stance that prevented the signing of a new treaty.

There are several alternative ideas that might replace Kyoto-style national limits on emissions. Intensity targets, a concept supported by many developing nations, are universal emission limits that are scaled to the size of a country's economy. Sector targets set limits for each industry, rather than each country. Programs that would allow nations to share emission control technology avoid binding targets altogether. "Under most of these systems of new, flexible targets," comments Bronwen Maddox in the London *Sunday Times*, "it might still be possible to set up markets in pollution, in which countries or industries could trade the right to release emissions. Any agreement to curb greenhouse gases is worth little if the US, China and India do not sign up. Kyoto failed in that basic requirement."[7]

Notes

1. James Kanter, "Per Capita Emissions Rising in China," *New York Times*, July 1, 2010. http://green.blogs.nytimes.com.
2. W. Kagi and H. Schmidtke, "Who Gets the Money? What Do Forest Owners in Developed Countries Expect from the Kyoto Protocol?" Food and Agriculture Organization of the United Nations, March 2005. www.fao.org.
3. John Vidal, "China Leads Accusation That Rich Nations are Trying to Sabotage Climate Treaty," *The Guardian*, October 5, 2009. www.guardian.co.uk.
4. Walter Russell Meade, "Why the World Is Better for Jesse Helms," The Jesse Helms Center, April 22, 2001. www.jessehelmscenter.org.
5. BBC.com, "U.S. Gases to Dwarf World Savings," March 7, 2005. http://news.bbc .co.uk.
6. James M. Taylor, "Europe to Russia: Ratify Kyoto or Else," *Environment and Climate News*, June 1, 2004. www.heartland.org.
7. Bronwen Maddox, "Why Kyoto Will Vanish into Hot Air," *The Times*, November 29, 2005. www.timesonline.co.uk.

CHAPTER 8

Carbon Trading

Companies set up factories, employ workers, and produce goods in order to earn a profit. In a capitalist economy, people are free to create businesses and partnerships to meet the same goal. But all business is regulated, to some extent, by government. Lawmakers can pass regulations and levy taxes on individual companies and on industrial sectors, such as energy and automaking. Government can also impose fines on companies that violate the law.

The need of business to earn money often is in conflict with the public's need to prevent toxic pollution and additional global warming. Whereas high taxes reduce economic activity, new streams of income improve company profits. By controlling greenhouse gas emissions with a cap-and-trade system, government environmental regulation imitates the free market in goods and services. This can give business a profitable incentive to lower greenhouse gas emissions.

The Cap-and-Trade Concept

The cap-and-trade concept is based on the premise that carbon dioxide released by human activities is the main culprit in global warming. In a cap-and-trade system, a government agency sets a limit or cap on the maximum amount of carbon dioxide that can be released by a single factory or power plant into the environment. The government grants emission permits, or carbon credits, which allow the plant to emit that amount and no more. If the factory needs to release a greater amount, it must buy more permits. If the factory releases less, it may sell the extra permits it holds on an open market, where such permits have a value that fluctuates with supply and demand. It may also sell the permits to other companies and factories that need them.

In theory, trading carbon credits will allow a government to reduce overall carbon dioxide emissions. If permits are taken out of circulation, or if their number is reduced by law, then their value and price rise. As a result, companies will—eventually—find it more economical to convert to "green" production, which will be cheaper than emitting the same amount of carbon dioxide and buying the permits.

Carbon trading represents an alternative to simple regulation. Although the government could, in theory, just limit the release of greenhouse gases by law (or tax it), such action would directly slow the economy. Instead of hiring workers and producing goods, companies would have a greater interest in cutting production to meet the law's requirements. As an alternative, cap and trade gives industries and individual companies several choices in the drive to reduce their carbon emissions.

Differing Approaches to Regulation

Since the 1960s, the Environmental Protection Agency (EPA) has tried several different approaches to environmental regulation. In the early years, the agency established guidelines and made the states, not individual industries, responsible for enforcing them. Most factories and power plants were free to operate as they wished, with little constraint on the emission of toxic waste and other harmful materials into the air and water.

Flexible regulation allowed the states the freedom to enforce the EPA's guidelines as they saw fit and industries the ability to meet state laws in different ways. This approach is still supported by industry leaders, who feel that environmental regulation should take local conditions into account. Nationwide mandates from the federal government, in this view, are inefficient and ineffective, and they create a heavy burden of costs both for private companies and public, tax-supported agencies.

Since the 1990s, the debate over environmental laws has intensified with the global warming issue and the Kyoto Protocol, an international agreement to limit carbon dioxide emissions that the United States has not joined. In one view, also known as the retail approach, regulation should be tailored to each industry and location. New emission rules should be subject to a cost-benefit analysis. If they cost private industry too much, and do not result in immediate environmental benefits, they should not be enacted. In addition, all regulation should be founded on peer-reviewed science, not on the whims of government bureaucrats. A majority of creditable scientists should support the underlying research that leads to new rules and regulations. In addition, companies should have the power to challenge the regulations, whether state or federal, in court.

The traditional approach, also known as command and control or end of the pipe regulation, allows government to set down strict standards and mandate the ways companies can meet them. There can be no challenge to the regulations by states or companies. The federal and state governments punish violators with fines and have the right to close down factories altogether. This approach results in more uniform standards across the country, but it also brings higher costs for companies that have limited options in meeting the standards. In addition, command and control is expensive to formulate and implement. One study by the Office of Management and Budget, a federal agency, found that passing and enforcing environmental regulations costs the United States government $150 billion a year.[1]

The Clean Air Act Amendments of 1977 and the Offset Mechanism

In 1970, Congress passed its National Ambient Air Quality Standards, or NAAQS. Under this law, the EPA had to set

What Is a Carbon Offset?

The carbon offset mechanism is a means of applying the free market to environmental regulation. The mechanism dates to the 1970s, when factories and companies were allowed to pay for their high-emission plants by operating others more cleanly. One carbon offset represents a specific amount of carbon dioxide (or other greenhouse gas). Companies buy these offsets on an open market, where the value fluctuates with supply and demand. By doing so, they support a renewable energy project in another location, which helps to reduce emissions and thereby offsets the production of greenhouse gases by the company. Carbon offsets can be used for solar, wind, or geothermal power projects, or for the reforestation of areas placed under cultivation. The Bonneville Environmental Foundation, for example, sells offsets that equal 1,400 pounds (635 kg) of greenhouse gas emissions; the organization claims they represent the replacement of fossil fuels with renewable sources producing 1,000 kilowatt-hours of electricity. One ClimateCare Verified Emission Reduction represents a savings of one metric ton of carbon dioxide not released into the atmosphere; similar offsets are offered by Conservation International and by the TerraPass Verified Emission Reductions.

standards for six pollutants: carbon monoxide, nitrogen dioxide, ozone, sulfur dioxide, lead, and particulate matter. The agency set primary standards to protect the health of the public, whereas secondary standards would enhance the general welfare by protecting forests, cropland, and the natural environment. Older factories were made exempt from regulation, as Congress believed they would gradually fall out of use as they grew obsolete.

In the following years many of the states did not meet the standards, however, and in 1977, Congress passed amendments to the Clean Air Act. Under the New Source Review provision, older plants were now required to undergo an EPA review whenever they sought to expand or renovate their facilities. These

Individuals can buy carbon offsets as well. Airlines, car rental companies, consumer-goods companies, and even Facebook sponsors offer these offsets to customers, promising to use the funds to support worthy environmental projects. The ultimate goal is to become carbon neutral, meaning all personal carbon production is balanced by the purchase of offsets and a reduction in carbon emissions somewhere else on the planet. Companies that produce and sell carbon offsets claim they are an effective way to combat greenhouse gas emissions. The Federal Trade Commission (FTC) has also stepped into the market with an investigation, however. Christopher Joyce quotes Jim Kohm of the FTC in "Carbon Offsets: Government Warns of Fraud Risks":

> Our concern is that because these claims are very hard to substantiate and consumers can't easily tell they're getting what they pay for, there is the real possibility of fraud in this market. . . . There's been an explosion in green marketing. There are claims that we didn't see in the market 10 years ago. Carbon offsets are one of those new claims.

plants also had to install pollution control equipment if their emissions were to exceed the 1970 standards.

The older factories had an option, however. They could offset any one of their older, dirty plants by lowering emissions at newer, clean plants. Even if a company expanded, it could offset its emissions in this way, and the net effect on the environment would be the same.

The 1990 Clean Air Act Amendments and the Acid Rain Program

The government had offered companies a flexible way of following the rules. In turn, individual companies could comply with the rules in the way that best served their interests. The regulations were largely successful, and in the interest of developing new market-based mechanisms for pollution control, the US Congress sponsored a study known as Project 88. This program allowed business leaders, lawmakers, and environmental groups input into future pollution control systems. Released in December 1988, Project 88 produced thirty-six ideas for solving environmental problems using an approach based on economic incentives, rather than direct regulation. It led directly to the cap-and-trade mechanism set up in the Acid Rain Program, part of the Clean Air Act amendments of 1990.[2]

Through the amendments, the offset mechanism was applied to sulfur dioxide and nitrogen oxides, the compounds that combine to form acid rain. The rules went into effect in two phases. Phase I ran from 1991 to 1995 and targeted major sources of acid rain pollutants. Phase II began in 1996 and applied to all fossil-fuel-burning electricity-generating plants.

These utilities had several options to meet the annual emissions targets: They could use cleaner-burning coal that had a lower sulfur content; they could buy emissions allowances for plants that did not meet the targets; or they could install gas-desulfurization scrubbers in their smokestacks. Eventually, the success of this flexible regulation inspired support for an offset mechanism for carbon dioxide emissions worldwide.

A direct tax has the virtue of simplicity: It makes older forms of energy and production more expensive, giving an immediate incentive to change.

Cap-and-trade mechanisms have also been established in states and cities. In the Chicago area, an Emissions Reduction Market System begun in 2000 allows major pollution sources to trade credits among themselves. Several states in the northeast have joined the Regional Greenhouse Gas Initiative, a trading system in carbon dioxide permits that began operating in 2009.

Carbon Allowances Versus Carbon Taxes

Instead of a global carbon-trading market, many economists still support carbon taxes as a more effective way of limiting carbon emissions. In this view, imposing such taxes will give companies and individuals a more direct incentive to reduce their carbon footprints. A direct tax has the virtue of simplicity: It makes older forms of energy and production more expensive, giving an immediate incentive to change. It can be altered from time to time as emissions levels rise and fall, and the government seeks to levy higher taxes on the most harmful emissions.

The effect of an energy or carbon tax on emissions, however, cannot be known in advance, so a tax has a less certain result than a trading program that limits overall emissions with permits. Taxes also have the effect of slowing down production and economic activity. If forced to pay taxes on their emissions, factories that produce carbon-intensive goods stand to lose income and profits, unless they can pass their higher costs on to customers. They will attract less investment and have less money available for new manufacturing or for making their operations more efficient.

Carbon trading, on the other hand, is a much more complex mechanism. Because there is a free market in carbon credits, no one can know their price in advance. Companies would be hard-pressed to start green-energy projects without knowing whether the carbon price would make such projects economical in the future.

In addition, setting emission targets is subject to the decisions of lawmakers, and trading carbon credits requires companies to divert resources to departments that carry out this task. Carbon trading can also invite cheating and corruption, as companies seeking to make a profit on the carbon market attempt to manipulate the price of credits in their favor. Government agencies, through their power to set the legal number of credits in circulation, may also act to benefit certain companies and not others.

Both systems carry uncertainty, either for the government agency trying to control emissions or for the industry attempting to make a profit. As new scientific data come in, and economic conditions change, the public may demand changes in the laws, bringing even more complexity to the system.

Carbon Leakage

Carbon dioxide and other greenhouse gases do not respect national borders. The effect of carbon emissions on the earth's atmosphere is the same no matter where the carbon originates, which gives rise to the leakage problem. If one nation is subject to limits on emissions, its industries may move to a neighboring country, where there are no limits on emissions either by domestic laws or international treaties. The result can be an increase in total emissions, if the second country has looser environmental regulations.

Leakage is most commonly a problem between two bordering countries that are at different levels of economic development. The United States and Mexico, for example, share a border running 2,000 miles (3,219 km) between the Pacific Ocean and the Gulf of Mexico. If a factory in the United States moves to Mexico and produces the same amount of goods, lax regulations in Mexico allow it to release more greenhouse gases, and the environmental impact is greater.

The Kyoto Protocol capped emissions on its Annex I countries, but the majority of nations did not fall into this category and therefore were not bound to reduce emissions within their borders. This has been a sore point for environmentalists, who see Kyoto as only a half measure that will have little effect in the drive to reduce emissions globally. The debate over any further treaties after Kyoto revolves around the inclusion of China and other countries in the treaty terms, in order to prevent leakage.

One method of solving this problem is through trade policy. Members of the United States Congress have proposed legisla-

A woman works in the US-owned Valmont Electric plant near the border city of Juarez, Mexico. Some companies relocate out of Annex I countries into countries with fewer environmental regulations. William F. Campbell/Time & Life Pictures/Getty Images.

tion known as the American Clean Energy and Security Act. If it is passed, this law will set up a national cap-and-trade market and place an import duty on goods produced in countries without a carbon-trading system. The tax would be set at a certain amount per ton of carbon dioxide released in order to produce the imported item. In theory, it will provide an incentive to other countries to set up carbon-trading systems of their own and adopt clean energy and other green technologies to reduce their carbon emissions.

Developing countries resist such taxes, however, as the "carbon duty" would make their own goods more expensive and less competitive. In this view, such carbon duties violate any trade treaties that exist between two countries that ban any further barriers to trade, and unfairly penalize workers in countries that, in the past, had no role in the problem of greenhouse gas emissions and global warming.

The Carbon Credit Market in Europe

The nations of Europe have signed and ratified Kyoto, and they are striving to meet the greenhouse gas limits placed on them by the treaty. To that end, the European Union (EU) set up an Emission Trading System (ETS), which began operating among the fifteen member nations in 2005. Since that time, the ETS has expanded to include twenty-five members of the EU. As the most important financial center in Europe, London has become the global center of carbon trading.

The ETS allows each country to set its own emission caps. The program limits carbon emissions by the continent's largest factories and power plants, defined as those using a certain amount of energy in their day-to-day operations. These plants are responsible for about half of the total carbon emissions throughout Europe. In the early phases of the program, governments softened the impact of carbon trading by setting relatively easy emissions limits, based on the emissions amount of previous years (a method known as grandfathering). In addition, many countries issued excess allowances, budgeting for more carbon emissions than were actually taking place. The result was a decline in the price and value of carbon credits.

Many companies in Europe have been earning extra income by selling and trading their allowances, and by setting up clean development mechanism (CDM) projects in developing nations. For a CDM project, a large international corporation will create a green-energy project for one of its own factories in a developing country, and in this way benefit from updating its plant, enjoy a savings in energy and labor costs, and earn carbon credits as well.

By most estimates, the EU trading program has been an overall success and has brought about a decline of about 3 percent in greenhouse gas emissions in 2008 alone.[3] The next phase of the

EU program is scheduled to begin in 2013, the year after Kyoto's deadline for emission reductions. To improve the system, the EU may set an overall cap on emissions for all of its members. It will also set a fixed number of allowances for the continent as a whole and allocate these allowances to each country according to that country's own carbon budget.

The Carbon Credit Market in the United States

Although the United States has not ratified the Kyoto treaty, there are several carbon-trading markets operating in the United States. American companies from any state can trade carbon credits on the Chicago Climate Exchange. In addition, coal-fired power plants are still trading sulfur dioxide allowances among themselves as part of the Acid Rain Program. The Western Climate Initiative began in 2007 and allows companies in seven states and four provinces of Canada to trade carbon credits.

Several states have set up a green-tag program, which rewards companies that are creating renewable energy. The source of power can vary, from wind and solar to geothermal energy and biomass conversion. A green tag is a credit for producing one kilowatt-hour of renewable energy. It can be sold from one company to another; the company purchasing the green tag can thus offset its use of fossil fuels to meet a voluntary cap on emissions that it has adopted.

The US government has taken action by allocating a part of its budget to clean energy projects. These projects can be rewarded with carbon credits that are issued by the government and can be sold by the company that earns them.

A mandatory cap-and-trade system in the United States is subject to heavy opposition from business leaders, however, and Congress has not passed the American Clean Energy and Security Act. Many opponents of carbon trading believe it represents a mandate that interferes with the operation of private business. Others who are skeptical of human-produced global warm-

THE FALLING CARBON CREDIT MARKET

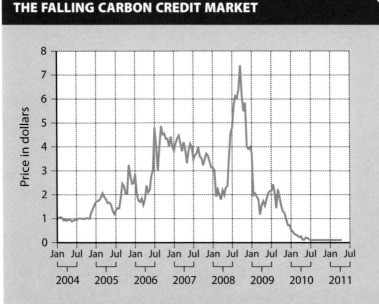

Since 2003, the daily price of carbon financial instruments, carbon credits (per ton), has fluctuated on the Chicago Climate Exchange (CCX), but reached a low of 10 cents per ton in the summer of 2010. In 2009, Congress proposed new legislation known as the Waxman-Markey American Clean Energy and Security Act, which would have required U.S. companies to take part in a carbon-trading system under federal law. To ease the transition, the law would have the government grant 90 percent of carbon allowances free to private companies between 2012 and 2026. With the proposal that most allowances be given away for free by a law that hasn't yet passed, carbon allowances fell to a price of 10 cents a ton, an all-time low.

Source: "CCX Carbon Financial Instrument (CFI) Contracts Daily Report." www.chicagoclimatex.com; "The Collapse of the CCX Carbon Emission Contract." http://seekingalpha.com.

ing see it as an expensive and ineffective idea that results only in higher costs to consumers, affects the competitiveness of individual companies, and does nothing to help the environment.

The rules and regulations governing carbon trading remain a patchwork among the fifty states. With several regional carbon markets operating, the United States has failed to create a national system that would help companies use carbon trading efficiently. In addition, the system requires a method of measurement and enforcement that is uniform across all the states; otherwise, the states will end up competing among themselves, and states with more lax standards will enjoy a competitive advantage over those with stricter laws and heavier fines.

Some environmentalists have come out against the idea of carbon trading altogether. Larry Lohmann of The Corner House, a Canadian environmental group, expressed his opinion that

> Carbon trading is aimed at the wrong target. It doesn't address global warming. Solving global warming means figuring out how to keep most remaining fossil fuels in the ground. It means reorganizing industrial societies' energy, transport and housing systems—starting today—so that they don't need coal, oil and gas. Carbon trading isn't directed at that goal. Instead, it's organized around keeping the wheels on the fossil fuel industry as long as possible.[4]

Lohmann and others also fear the possibility of companies evading or misleading regulators, or of buying carbon credits from polluters in developing nations, where standards for carbon emissions may not be present at all. In addition, opponents see the potential for cheating by companies that plan CDM projects in developing nations and claim a savings in emissions where little or none exists. Opponents of carbon trading also believe that it distracts from the real challenges of developing alternative energy sources. Carbon taxes, in this view, are much simpler to understand, less subject to deception, and easier to enforce on the part of public agencies.

By one estimate, $64 billion worth of carbon credits were traded on carbon exchanges around the world in 2007, a sharp increase from the volume of the previous year.[5] There is an active carbon market in Europe, where carbon emissions limits have been effectively imposed since the passage of the Kyoto treaty. In the United States, however, where future law is uncertain and carbon emissions are still not uniformly regulated by law, the carbon market is dormant, and companies are channeling their environmental activism to innovation in their processes and product design, rather than through the cap-and-trade idea.

Notes

1. Robert W. Hahn and Fumie Yokota, "Regulation," in *The Oxford Companion to Global Change*, New York: Oxford University Press, 2009, p. 527.
2. "Project 88—Round II: Incentives for Action: Designing Market-Based Environmental Strategies," May 1991. www.hks.harvard.edu.
3. "Emissions Trading: EU ETS Emissions Fall 3% in 2008," *Europa*, May 15, 2009. http://europa.eu.
4. Larry Lohmann, "Six Arguments Against Carbon Trading," Climate and Capitalism, September 29, 2008. http://climateandcapitalism.com.
5. Green Living Online, "Canada Enters Carbon Exchange Market with Montreal Climate Exchange Launch." www.greenlivingonline.com.

CHAPTER 9

Green Industry

Foreign competition is a vital consideration for any business. Lower labor and material costs in other countries may allow competitors to sell products at lower prices. This advantage helps competitors take market share away from companies that must meet higher expenses. Regulation of industry is an important part of this equation. Government control of greenhouse gas emissions means that many companies must invest in clean energy projects to stay within mandates set down by the authorities.

There is an upside to the expense of green technology, however, because a vast new marketplace is evolving. Green products and services are needed in any country that is striving to meet limitations imposed by the Kyoto Protocol—which went into effect in 2005 to limit carbon dioxide emissions of more than fifty nations—and other agreements. Companies that produce green products and technologies can thrive from meeting this demand.

Advocates of emission control point out that without government controls on greenhouse gases, companies have no motivation to invest in clean energy. Without developing this technology for their own benefit, they are at a disadvantage in foreign markets that are rapidly developing green industries.

As far as consumers are concerned, cost savings is the most important motivation for "going green" and for buying products designed to reduce or eliminate greenhouse gas emissions. One study carried out by researchers found that 26 percent of consumers buy green products to lessen their impact on the environment; 73 percent use them to save on their household bills.[1]

Public Image: Promotion and Proaction

In business, where companies are in the constant process of creating a brand, public image is important. A brand is much more than a slogan or familiar logo; it is the entire set of impressions that the public has of a company and its products. Environmentally friendly brands draw more support and easier public acceptance of their products. Industries that strive for green production methods also gain credit with the public for helping to resolve environmental issues such as pollution and global warming.

As lawmakers pass stricter mandates and controls, businesses either change their methods or face a problem that becomes much larger, and more damaging, than poor brand image. A limit on carbon emissions may force a manufacturing plant to slow its production if the plant has not already prepared by adopting green technology, recycling its waste, or bringing renewable energy sources online. Sudden conversion under the threat of fines or other penalties ends up costing more than would preparation for the conversion ahead of time.

Business leaders, by and large, still oppose mandatory restrictions on the way they operate and how they use raw materials and energy. Most favor voluntary reductions in carbon emissions, which can be done at the pace that is most appropriate to each individual company and industry. In the United States, which is not a party to the Kyoto treaty, the voluntary approach is still favored. In Europe and developed nations of Asia, greenhouse gas emission limits are set by governments, often in the form of emission allowances, and enforced with fines and sanctions. By 2010 Europe and Japan had achieved more success in limiting emissions than had the United States—but their efforts had come at greater cost to their home industries.

GREATEST INFLUENCERS ON AN ORGANIZATION'S ENVIRONMENTAL BEHAVIOR

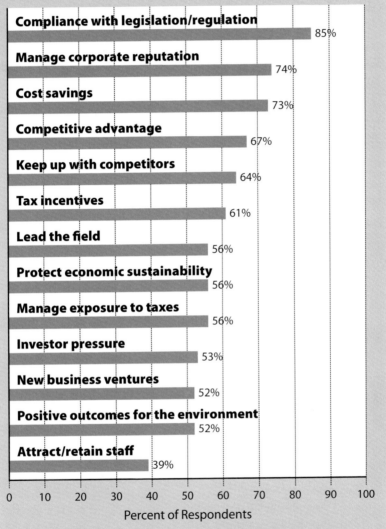

Compliance with legislation/regulation — 85%

Manage corporate reputation — 74%

Cost savings — 73%

Competitive advantage — 67%

Keep up with competitors — 64%

Tax incentives — 61%

Lead the field — 56%

Protect economic sustainability — 56%

Manage exposure to taxes — 56%

Investor pressure — 53%

New business ventures — 52%

Positive outcomes for the environment — 52%

Attract/retain staff — 39%

Percent of Respondents

Based on nearly 700 interviews conducted in businesses worldwide with the "person in the organisation responsible for setting company policy and strategy towards managing its environmental impact and the main costs associated with this."

Source: PriceWaterhouseCoopers, "Appetite for Change," May 2010. www.pwc.com.

Minimizing Impact Through Green Design

Green technology attempts to minimize the environmental impact of production, which means reducing emissions by installing pollution control equipment and using new production methods. This lowers the amount of toxins and greenhouse gases released into the environment. Factories can also switch to renewable or recycled resources as raw materials for their finished products. Many companies are now touting their use of recycled materials in packaging, such as food and beverage containers, which in many sectors makes up a significant cost of business.

To minimize the impact of resource use, some companies are taking part in replenishment programs in which, for example, trees are planted to match the depletion of forests caused by the creation of paper or wood pulp. Other companies are using offsets, which means financing a project, either at home or abroad,

The Green Seal of Approval

When manufacturers set out to make an environmentally friendly product, they seek an official seal of approval—some certification that indicates to purchasers that their new item meets certain strict environmental guidelines. The nonprofit organization Green Seal was founded in 1989 to establish environmental and health standards for consumer products. Manufacturers submit the ingredients and manufacturing process to the organization. If the product passes the standards, the Green Seal logo appears on the packaging. Green Seal standards and labeling have been adopted by hotels, food companies, government agencies, and building contractors.

Another certification program is Cradle to Cradle, operated by McDonough Braungart Design Chemistry. New products submitted to McDonough Braungart must come with a complete disclosure of all chemicals that go into the finished product. The company is looking for completely recyclable materials—such as nylon, steel, polypropylene, that neutralizes the environmental impact of manufacturing and other operations by the company at large.

Carbon Footprinting

Whereas toxic pollution from waste pipes and smokestacks once labeled a factory as hostile to the environment, the emission of carbon dioxide and other greenhouse gases now stands as the leading measure of environmental impact. With this new idea, the concept of the carbon footprint has emerged. A large carbon footprint means a plant or business emits an excessive amount of carbon dioxide into the atmosphere. A zero footprint means no net release of carbon into the environment.

One major company now promotes its manufacturing processes as carbon negative. Milliken and Company, based in South Carolina, once produced carpets and textiles using manufacturing methods that had a high environmental impact. This

and aluminum—and nontoxic materials. The first gold certificate awarded through the program was for the Zody, an office chair that was launched in 2005 by the Haworth office product company. The Zody is 98 percent recyclable and can be disassembled in about 15 minutes with common hand tools. After the Zody appeared, competitors rushed to market with their own eco-friendly office chairs, including the Think chair made by Steelcase and the Mirra, made by Herman Miller.

As a result of the Cradle to Cradle program, furniture makers are now minimizing the connecting parts on their items and replacing chrome with stainless steel. Powder coatings have replaced solvent-coated metals, and polypropylene and urethane have replaced polyvinyl chloride (PVC). The chairs are shipped in reusable blankets, not cardboard, and with backs separated from bases (which lowers shipping costs as more units can be loaded onto a single truck). Haworth even offers a take-back program, which allows owners to return their used chairs to the company for a recycling credit.

company has added more than two hundred environmental engineers to its payroll and has transformed its production process to reduce energy use and water consumption. The company has built its own hydroelectric plants to power its factories, and it has achieved a zero-waste target among its carpet plants, which send nothing to public landfills.

Milliken has planted and manages more than 100,000 acres of forests designed to serve as carbon sinks to offset the company's own greenhouse gas emissions. The net result is that the company draws more carbon dioxide out of the environment than it puts in—it has become a carbon sink of its own. The company benefits from an environmentally friendly image, but it also stands to profit from the sale of any carbon credits it earns through its carbon-negative processes.

Another green-conscious corporation is Ikea, a Swedish firm that operates large retail outlets selling furniture and home fixtures all over the world. Like many other companies, Ikea is seeking to develop an environmentally friendly brand. One way it does so is by charging customers five cents for each disposable plastic bag used and donating the proceeds to American Forests, a nonprofit corporation that plants trees in order to offset carbon emissions created by donor companies. Ikea and many other stores are also offering totes (made from recycled materials) that customers can bring back and reuse for carrying their goods. One study published by the Inhabitat website claims that

> 71% of all IKEA products are recyclable. . . . When a country introduces stricter emissions rules, like when Japan decided to restrict formaldehyde emissions to levels close to zero, IKEA imposes the new restrictions on its global operations. As a result, Ikea's policy reflects the strictest emissions policies in countries across the world, even though it sometimes drives costs higher.[2]

Following page: By charging customers for plastic bags, Ikea has reduced its carbon footprint and promoted a public image of environmental friendliness. AP Images/George Widman.

New Technologies and Materials

New technologies are giving rise to new industries. Many industrial engineers are now working in the field of nanotechnology, in which products are built at the size and complexity of a human cell. Scientists have proposed dozens of useful applications for nanotechnology: Microscopic machines may be able to regulate or repair body functions, generate useful energy by the conversion of sunlight, or serve as manufacturing robots for semiconductor companies. Nano products do not use fossil-fuel energy, and their use leaves no carbon footprint.

Nanotech and other green industries will require new forms of industrial materials; at the same time, old forms of industrial raw material may be phased out as public agencies pass stricter carbon limits. One of these materials is polyvinyl chloride, or PVC, a hard plastic that currently has thousands of uses but also carries a reputation as environmentally unfriendly. The complex manufacturing process of PVC creates polluting waste products as well as greenhouse gas emissions; in addition, PVC and other plastic products do not break down in the environment once they are discarded. PVC is useful and inexpensive, however, and more than 14 billion pounds of it are made every year in North America alone.

During the 1990s, the use of PVC began to affect the brand of companies such as Apple, a high-tech giant. Apple took a proactive stance, setting down strict guidelines for manufacturers that supplied it with parts for its laptop and desktop computers, and for assemblers who put the machines together. The company banned the use of chlorine and bromine entirely, and it also helped to develop a new plastic resin to replace PVC used in computer and printer cables. The company's engineers visit the manufacturing plants in China where Apple computers are assembled and interview managers, as well as randomly selected workers, in order to make sure its guidelines are being met. Although the initial expense of this process was high, the company earned praise from environmental groups for its efforts, as well as broader public acceptance of its products. Apple also asks its suppliers to agree to a Supplier Code of Conduct, which says, in the introduction, that it "outlines a comprehensive set of expectations covering labor, human rights, the environment, health and safety, ethics, and management systems. Every Apple supplier contract includes a commitment to comply with the Code."[3]

As new products and materials gain acceptance in the marketplace, the costs of manufacturing them typically come down.

Making and using PVC has health as well as environmental hazards. The basic raw material in PVC is vinyl chloride, which is carcinogenic. The manufacturing process gives off dioxin, a toxic chemical that is also released when PVC is burned. Toxic compounds are also added to some forms of PVC to give it greater strength and flexibility. Yet manufacturers face a serious challenge when trying to come up with a feasible alternative for PVC water pipes, soda bottles, product packaging, construction material, and toys. One company, Steelcase, has devised a nylon replacement for PVC-coated electrical wire and has phased out all PVC used in its office furniture and cabinets. But replacements must have all the useful qualities of PVC, including strength, resistance to water damage, and the ability to carry printing and embossing.

Distribution Alternatives

As new products and materials gain acceptance in the marketplace, the costs of manufacturing them typically come down. Economies of scale are created when large amounts of anything are made in a centralized manner; material and labor costs for each unit come down. One expense of manufacturing that consumers rarely confront is the transportation of that product from the factory to the store. Distribution—getting products to market—involves the use of energy, for the most part generated by fossil fuels, to power planes, ships, trains, and trucks.

Some companies are switching to hybrid vehicles to cut down on fossil-fuel use. Others are devising new ways of packaging goods for transport. Instead of using boxes to transport heavy furniture, for example, Allsteel and other companies now wrap their goods in reusable blankets and recyclable polybags. This method reduces expenses and increases the available space in the company's trucks. When fewer trucks are needed to transport goods, the company's carbon footprint shrinks even further.

A movement to make and sell products on a local, rather than international scale, has also taken hold. Locally produced goods have the benefit of low transportation cost, as well as supporting job creation and economic health closer to home. The movement runs across a wide range of basic household needs: food, furniture, clothing, tools, utensils, and many other products that are in common, constant demand.

The local-food movement has given rise to a new term: locavores, who make it a point to consume food produced within 100 miles of their home. Local production is closely tied to the concept of sustainable agriculture, the drive to reduce the impact of farming and food processing on the environment. According to the website Sustainabletable.org:

> By adding transportation, processing and packaging to the food system equation, the fossil fuel and energy use of our current food system puts tremendous stress on the environment. For example, between production and transportation, growing 10% more produce for local consumption in Iowa would result in an annual savings ranging from 280,000 to 346,000 gallons of fuel, and an annual reduction in CO_2 emissions ranging from 6.7 to 7.9 million pounds.[4]

Consumer Action

Companies seeking a positive public image are marketing green consumerism as well as green products. The lighting industry, for example, is benefiting from environmentally conscious consumerism. Philips, GE, and other companies are manufacturing compact fluorescent light (CFL) bulbs that consume less energy, and last longer, than conventional incandescent bulbs. The CFL has become an international emblem of successful new green products.

CFLs are gaining widespread support, despite their higher initial cost, due to their promise of reducing greenhouse gases and, in the long run, saving money for users. Australia and other nations favor CFL use and have passed laws banning traditional incandescent light bulbs in public facilities. Similar laws are being proposed or have already passed in California and New Jersey.

The widespread adoption of CFLs does not represent an easy solution to the global warming issue, however. CFLs contain the toxic element mercury. Disposing of used CFLs in public landfills eventually releases that mercury into the environment as the bulbs are broken and the mercury inside is released. According to a 2008 report of the EPA's Energy Star program, CFLs are a minor source of mercury; electricity generation accounts for the majority, in the amount of 104 tons of mercury released into the atmosphere every year: "If all 290 million CFLs sold in 2007 were sent to a landfill (versus recycled, as a worst case)—they would add 0.16 metric tons, or 0.16 percent, to US mercury emissions caused by humans."[5] Most states require that consumers recycle these bulbs by separating them from ordinary household waste and turning them in to a recycling program. Ikea and other companies have established take-back programs that offer CFL-recycling bins within their stores.

Sling Media, makers of the Slingbox, incorporated several green features into the design of the Slingbox 700U, a device that allows users to draw video content remotely from satellite and online providers. The new version of the product is small—about the size of a checkbook—and wrapped in a frame made of sturdy aluminum, making bulky packaging unnecessary. Method Products, which labels itself as an eco-friendly consumer products company, has also cut down on package size. The company has developed a new family of laundry detergents that buyers dispense from small pump-action bottles. The product is extremely concentrated, which cuts down on packaging, carbon use, and waste.

Sling Media, Method Products, and hundreds of other companies are entering the green market, for the sake of a positive brand image and, naturally, to make a profit. Yet as claims of environmentally friendly products and green production on the part of businesses multiply, consumer skepticism rises as well. The public realizes that these claims can be exaggerated, and what's labeled "green" may do as much harm to the environment as ordinary products that make no such claim. In addition, many green products don't compete well against traditionally made products that are less expensive and work just as well. Widespread acceptance and successful competition against environmentally harmful products are the keys to the future growth of a green consumer economy.

Notes

1. GreenBiz.com, "The Simple Secrets to Successful Green Marketing." www.greenbiz
   .com.
2. Adrianne Jeffries, "Is it Green? Ikea," Inhabitat, January 29, 2009. www.inhabitat.com.
3. "Supplier Responsibility 2009 Progress Report." http://images.apple.com.
4. "What is Local?" Sustainable Table, January 2009. www.sustainabletable.org.
5. EnergyStar.gov, "Frequently Asked Questions: Information on Compact Fluorescent Light Bulbs (CFLs) and Mercury," Energy Star, November, 2010. www.energystar.gov.

CHAPTER 10

Conclusion

In the matter of global warming, as in other environmental issues, industry has often stood squarely against government. In the United States this contest dates to the 1960s and the first guidelines on air pollution set by the federal government. Although regulations on leaded gasoline, sulfur dioxide emissions, and CFCs eventually solved serious environmental problems, these rules did not enjoy much support among business leaders when they were first proposed. The same is true in the twenty-first century, when rising atmospheric levels of carbon dioxide and other greenhouse gases threaten a global environmental crisis.

Many business leaders believe that new regulations on carbon emissions will raise their costs and make them less competitive with foreign companies that do not have to operate within the same rules. Although the government supports carbon trading as an efficient, free-market solution, business owners know that hiring people and systems to handle the carbon market will cost money; the costs will be passed along to customers, and the overall economy will suffer. The retooling of assembly lines and manufacturing plants to conform to new standards also requires investment, as does the creation of green products to serve a new market.

In the meantime, scientists largely agree on the anthropogenic origins of the global warming that has been measured in recent decades. With clear evidence in hand of rising temperatures correlated to rising carbon dioxide levels, most environmentalists are calling for a global top-down solution to the issue because the effects of global warming and climate change do not respect national borders. The Kyoto Protocol, an international agreement to limit carbon emissions, is an example of one such approach: It establishes very specific limits on individual nations even as it allows those nations the flexibility to meet the limits in several different ways.

Others believe in a flexible approach that takes into account the needs of individual countries and the capabilities of individual companies. The United States and western European countries are supporting the more flexible bottom-up approach for the next global environmental agreement, the successor to the Kyoto Protocol. Developing nations strongly oppose the adoption of flexible regulation by the developed world, however, and feel that the United States and Europe are trying to relieve themselves of the major responsibility of solving the problem that they largely created. The basic debate over the best method of environmental regulation remains a contentious issue.

Nevertheless, business leaders are attending to their work and planning for new streams of income and profits. Many different industries have opportunities in the growing green market for environmentally sound goods. Companies able to produce items that save on the use of fossil fuels and carbon emissions will profit, as long as they can compete with businesses in other nations. Because transportation costs have come down, nearly all green products have a ready market in far-flung corners of the globe, including many places, such as Europe, where environmental standards are strictly enforced by governments.

The global warming issue will continue to make headlines. Business and government leaders all have a common interest in taking action, for the sake of either consumer acceptance or for votes. The goal for both factions is simple: a solution to the problem of global warming, along with an expanding and profitable marketplace for goods.

Glossary

acid rain The result of sulfur dioxide and other compounds mixing with precipitation in the atmosphere.

allowances Permits that allow a company or single plant to emit a specific amount of carbon dioxide or other greenhouse gas.

anthropogenic Created by humans or human activity.

biofuel Energy-rich material used to power machinery and made from biodegradable sources, such as crops or vegetable oils.

CAFE standards Corporate Average Fuel Economy standards, which are guidelines for vehicle manufacturers in setting minimum fuel mileage.

cap and trade A system of limiting greenhouse gas emissions and issuing permits for those emissions that can be traded on an open market.

carbon capture and storage A method of capturing carbon dioxide before it is released into the atmosphere and of placing it in a permanent facility or natural formation.

carbon footprint The amount of carbon dioxide released into the atmosphere by a single activity, facility, company, or individual.

carbon offset A credit that can be purchased by a company or individual and invested in a clean energy project to counterbalance the buyer's own carbon production.

carbon sequestration Storing carbon dioxide in a permanent facility or natural formation.

carbon sink A natural feature that absorbs carbon dioxide from the atmosphere.

chlorofluorocarbons Artificial chemical compounds that degrade the protective ozone layer in the earth's upper atmosphere.

clean development mechanism A system of investment in environmentally friendly energy and resource projects in developing nations, thus limiting use of fossil fuels and the release of greenhouse gases into the atmosphere.

command and control regulation A method of limiting toxic pollution and greenhouse gas emissions by establishing laws, guidelines, and regulations.

feed-in tariff Money paid to a private producer of electricity, who can sell unneeded power to a utility company.

flexible regulation Allowing local governments or nations to set emission limits that are appropriate for their own economic conditions and stage of development.

geothermal energy Power produced by drawing on heat beneath the earth's surface.

greenhouse effect A warming of the atmosphere caused by the trapping of sunlight by gases such as methane, water vapor, and carbon dioxide.

hybrid A vehicle running on two or more sources of energy, such as conventional gasoline fuel and electricity.

Keeling curve A chart, created by Charles Keeling during the 1950s, that shows the gradual rise of carbon dioxide concentrations in the atmosphere.

locavore An individual who consumes food produced in the vicinity of his or her home.

mandate A regulation or rule imposed by government upon a company or community.

ozone A molecule that, when present in the stratosphere, helps to block the sun's harmful ultraviolet rays.

particulates Dust and soot, emitted by industrial sources, that collect in the atmosphere and eventually return to the earth.

photovoltaic cell A device for transforming sunlight into electrical current.

renewable sources Reservoirs of energy that occur and are replenished naturally, such as wind and sunlight.

smog A combination of chemical compounds created by factory and vehicle emissions that create a brown haze over large cities.

sustainable agriculture Farming that draws on renewable sources of energy for the production of food.

For Further Research

Books

Ray C. Anderson with Robin Wright, *Confessions of a Radical Industrialist: Profits, People, Purpose—Doing Business by Respecting the Earth*. New York: St. Martin's, 2009.
> Ray Anderson describes his effort to make Interface, his carpetmaking company, the first in the nation to achieve 100 percent sustainability with minimal impact on the local and global environments.

Janine M. Benyus, *Biomimicry: Innovation Inspired by Nature*. New York: Harper Perennial, 2002.
> Pointing out that humans are the most wasteful and energy-intensive species on the planet, the author offers solutions in design that mimick the more efficient mechanisms found in the natural world.

Lester Brown, *Plan B 4.0: Mobilizing to Save Civilization*. New York: Norton, 2009.
> An exploration of alternative energy sources and the feasibility of using them to completely replace fossil-fuel use by individuals and organizations.

Robert Bryce, *Power Hungry: The Myths of "Green" Energy and the Real Fuels of the Future*. New York: PublicAffairs, 2010.
> Believing that renewable energy sources are still inefficient and inadequate to meet the needs of modern societies, the author suggests natural gas and nuclear power as more effective and environmentally clean solutions.

Helen Caldicott, *Nuclear Power Is Not the Answer*. New York: New Press, 2007.
> The author points out the issues still facing nuclear power, including inadequate storage of waste, environmental impact of nuclear stations, fossil-fuel emissions in the production and construction of nuclear facilities, and huge start-up costs.

Gwyneth Cravens, *Power to Save the World: The Truth About Nuclear Energy*. New York: Vintage, 2008.

The case for nuclear energy as a relatively safe and environmentally sound alternative to the use of traditional fossil fuels, including coal, natural gas, and oil.

Andres R. Edwards, *The Sustainability Revolution: Portrait of a Paradigm Shift*. New Society, 2005.
> A description of the "sustainability revolution," as practiced in various countries and by various companies.

Daniel Esty and Andrew Winston, *Green to Gold: How Smart Companies Use Environmental Strategy to Innovate, Create Value, and Build Competitive Advantage*. New Haven, CT: Yale University Press, 2006.
> A book describing the environmental policies of various companies and how businesses can benefit financially from green policies, products, and image.

Thomas Friedman, *Hot, Flat, and Crowded: Why We Need a Green Revolution—And How It Can Renew America*. New York: Farrar, Straus and Giroux, 2008.
> The author calls for government policy to address the global warming issue and believes the United States should make a fundamental change to sustainable industry, energy production, and agriculture.

David Gottfried, *Greed to Green: The Transformation of an Industry and a Life*. Berkeley, CA: WorldBuild, 2004.
> A former real-estate developer explains how businesses can make a profitable transition to sustainable production, and as well as how environmentally friendly construction methods make for healthier work and living places.

Paul Hawken, Amory Lovins, and L. Hunter Lovins, *Natural Capitalism: Creating the Next Industrial Revolution*. New York: Back Bay Books, 2008.
> The authors foresee a transformation in manufacturing and product design that will lead to faster economic growth and a more abundant way of life.

Richard Heinberg, *The Party's Over: Oil, War, and the Fate of Industrial Societies*. Gabriola Isand, BC, Canada: New Society, 2005.

The author warns of the coming "peak-oil" event that will bring about increasing cost and scarcity of this resource and spark conflict among the nations over its control and use.

James Howard Kunstler, *The Long Emergency: Surviving the End of Oil, Climate Change, and Other Converging Catastrophes of the Twenty-First Century.* New York: Grove, 2006.
 The coming oil and climate crisis will effectively destroy the suburban car culture of the United States, in the author's opinion, and eventually bring about a new social model, in which small communities must draw on local food and resources for their survival.

William McDonough and Michael Braungart, *Cradle to Cradle: Remaking the Way We Make Things.* San Francisco: North Point Press, 2002.
 The authors propose new methods of production and product management that would eliminate waste and lead to increased efficiency and minimal environmental impact.

William J. Mitchell, Christopher E. Borroni-Bird, and Lawrence D. Burns, *Reinventing the Automobile: Personal Urban Mobility for the 21st Century.* Cambridge, MA: MIT Press, 2010.
 A book presenting a plan for redesigned urban transportation systems and a new, environmentally sound approach to automaking.

Eric Poole, *The Climate War: True Believers, Power Brokers, and the Fight to Save Earth.* New York: Hyperion, 2010.
 An exploration of the climate debate and how powerful special interests have successfully blocked any effective action on carbon emissions by the government.

Paul Roberts, *The End of Oil: On the Edge of a Perilous New World.* Boston: Mariner Books, 2005.
 The author describes the effects of oil depletion on the world economy and modern industrialized societies, as well as on developing countries such as China and India that are reliant on fossil fuels for their rapid economic growth.

Daniel Sperling, *Two Billion Cars: Driving Toward Sustainability.* New York: Oxford University Press, 2010.

A review of the modern car industry and the possible future of a car-dependent society, with ideas on how transportation systems and methods will have to change in the future.

Nicholas Stern, *The Global Deal: Climate Change and the Creation of a New Era of Progress and Prosperity*. New York: PublicAffairs, 2009.
A leading economist analyzes the costs of global warming and the need for rich and poor nations to strike a deal to reduce carbon emissions.

Periodicals and Internet Sources

Robert D. Atkinson and Daniel D. Castro, "Tomorrowland: In the City of the Future, Bridges Will Talk to Engineers, Roads Will Control Cars, and Parking Spots Will Find You. In Some Places, It's Already Here," *Atlantic Monthly*, June 2010, p. 64.

John M. Broder, "Obama to Face New Foes in Global Warming Fight," *New York Times*, November 3, 2010.

Ronald Brownstein, "The California Experiment," *Atlantic Monthly*, October 2009.

Center for American Progress, "Excuses, Excuses: Ten Industry Arguments Against Action on Global Warming . . . And Why They Are Wrong," May 30, 2008. www.american progress.org.

Center for Media and Democracy, "The Airline Industry's Global Warming Denial," December 1, 2009. www.prwatch .org.

Jonathan Chait, "E.P. Yay: Only Bureaucrats Can Solve Global Warming," *New Republic*, July 8, 2010, p. 2.

Tim Dickinson, "Climate Bill, R.I.P.: Instead of Taking the Fight to Big Polluters, President Barack Obama Has Put Global Warming on the Back Burner," *Rolling Stone*, August 5, 2010, p. 41.

Environmental Defense Fund, "Blueprint Lays Out Clear Path for Climate Action," May 8, 2007. www.edf.org.

Greenbiz, "Sustainability Still Wins Elections," November 3, 2010. www.greenbiz.com.

Joshua Hammer, "The $15 Trillion Treasure at the End of the World," *Fast Company*, November 2010.

Inhabitat, "Is It Green? Ikea," January 29, 2009. www.inhabitat .com.

Clifford Krauss and Jad Mouawad, "Oil Industry Execs Ready to Work on Global Warming," *New York Times*, February 11, 2009.

Joshua Kucera, "Side by Side in Need for Green Growth: China and America Try Cooperation," *U.S. News & World Report*, April 1, 2010, p. 42.

Steve Nadis, "SubTropolis, U.S.A.: A Large Chunk of Kansas City's Real Estate Lies 100 Feet Below Ground, and Offers a Creative Solution to Global Warming," *Atlantic Monthly*, May 2010, p. 20.

Kyunghee Park, "Big Ships Go Green," *BusinessWeek*, May 17, 2010, p. 35.

John H. Richardson, "Don't Need a Weatherman to Know Which Way the Wind Blows," *Esquire*, April 2010, p. 100.

Camille Rickets, "Making Some Eco Bets," *Fast Company*, July/ August 2010.

Mark Schapiro, "Conning the Climate: Inside the Carbon-Trading Shell Game," *Harper's*, February 2010, p. 31.

Scientific Activist, "Industry's Anti-Global Warming Misinformation Campaign Reminiscent of Big Tobacco's Strategy," April 24, 2009. http://scienceblogs.com.

Allan Sloan, "If You Believe in Magic, Green Energy Will Be Our Salvation," *Fortune*, July 26, 2010, p. 59.

Jeff Swartz, "Timberland's CEO on Standing up to 65,000 Angry Activists," *Harvard Business Review*, September 2010.

TerraDaily, "Global Warming Expert Raises Concerns for Tourism Industry," April 29, 2008. www.terradaily.com.

Clive Thompson, "Disaster Capitalism: Is the Planet Really Warming Up? Just Ask the Corporations That Stand to Make—or Lose—Billions Due to 'Climate Exposure,'" *Mother Jones*, July/August 2010, p. 32.

Web Sites

Environmental Defense Fund (www.edf.org). An organization that supports new laws, regulations, and business practices designed to protect the environment and combat anthropogenic climate change.

Environmental Protection Agency (www.epa.gov). This federal agency is responsible for enforcing federal laws on air quality and is the lead agency in implementing federal programs concerning global warming and climate change.

Intergovernmental Panel on Climate Change (www.ipcc.ch). This scientific association, organized by the United Nations, publishes authoritative research on greenhouse gas concentrations, global warming, and climate change.

International Energy Agency (http://iea.org). An organization with an interest in energy policy and innovation around the world, and which provides information on new energy systems currently under development.

National Oceanic and Atmospheric Administration (www.noaa .gov). This federal agency is responsible for scientific research and the collection of important data on the state of the ocean and the earth's climate.

Pew Center on Global Climate Change (www.pewclimate.org). A private think tank that addresses global warming, climate science, energy efficiency, and government initiatives on climate and green technology.

Real Climate (www.realclimate.org). A commentary site on climate science and other environmental topics, with blogs and articles written by working climate scientists.

Sierra Club (www.sierraclub.org). An environmentalist group that advocates for protection of natural resources and action by private industry and public agencies on global warming.

Index

About the Author

Tom Streissguth is the author of more than one hundred history, reference, and other nonfiction books for young people. He specializes in biography, history, geography, and current-affairs topics. A graduate of Yale University, he has worked as an editor, teacher, and paralegal. He currently lives in Minnesota.